Sustainability through Soccer

Sustainability through Soccer

An Unexpected Approach to Saving Our World

Leidy Klotz

UNIVERSITY OF CALIFORNIA PRESS

University of California Press, one of the most distinguished university presses in the United States, enriches lives around the world by advancing scholarship in the humanities, social sciences, and natural sciences. Its activities are supported by the UC Press Foundation and by philanthropic contributions from individuals and institutions. For more information, visit www.ucpress.edu.

University of California Press
Oakland, California

Library of Congress Cataloging-in-Publication Data

Names: Klotz, Leidy, 1978– author.
Title: Sustainability through soccer : an unexpected approach to saving our world / Leidy Klotz.
Description: Oakland, California : University of California Press, [2016] | Includes bibliographical references and index.
Identifiers: LCCN 2016009423 | ISBN 9780520287808 (cloth : alk. paper) | ISBN 9780520287815 (pbk. : alk. paper) | ISBN 9780520962859 (ebook)
Subjects: LCSH: Sustainability. | Environmental responsibility. | Soccer--Social aspects.
Classification: LCC GE195.7 .K586 2016 | DDC 338.9/27—dc23
LC record available at http://lccn.loc.gov/2016009423

Manufactured in the United States of America

25 24 23 22 21 20 19 18 17 16
10 9 8 7 6 5 4 3 2 1

To students young and old—
who inspire my endless quest

Contents

Acknowledgments xi

1. BACKGROUND 1

On sustainability, what soccer and systems-thinking have to do with it, and how I'm qualified to explain.

My Delayed Epiphany about Sustainability 1

Why Soccer? Sócrates Has Our Answer 3

The Black Panther and the Sacred Monster—Systems-Thinking for Sustainability 5

Why Me? An Autobiography in Less Than Twelve Hundred Words 9

How to Use This Book 13

2. PARTS 15

Systems-thinking is the guide for our sustainability quest, and systems are made up of some basic parts: elements, flows, stocks, feedback loops, and purposes.

Maputo and the Sacred Monster—Overview: System Parts 15

People and Players—System Elements 18

Floating Jabulanis and the River of Grass—Overlooked System Elements 20
"Remember Istanbul" and the Cross Bronx Expressway—Physical Flows 25
Smoking Managers and Electric-Bill Peer Pressure—Information Flows 30
Porto/Chelsea and a Russian "Oil"garch—Stocks 33
Obsessive Shooting Practice and Population Bombs—Feedback Loops 37
Falling Balls in Baltimore and Happy Bhutanese—Purposes 40
Review 43

3. BOUNDARIES 44

Our approach to sustainability depends on how we define our system boundaries. What must we consider? What can we leave out? And what can we reasonably infer from the perspective we define?

From Droughts to Floods in Maputo and Captain Sacred Monster—Overview: System Boundaries 44
"El Loco" Higuita and the Nine-Dots Puzzle—Space Boundaries 46
Glory-Days, Sour Grapes, and Seven Generations—Time Frames 50
A Bent-Legged Angel and Sustainable Hamburgers—Level of Detail 53
Icelandic Volcanoes and the Best Team Ever—Inputs and Outputs 59
Greece vs. Europe and Ehrlich vs. Simon—Insight, Not Clairvoyance 61
Review 65

4. BEHAVIORS 66

With parts and boundaries defined, we can detect system-level behaviors and discover how they affect sustainability.

A United Nations Report and the Sacred Monster's Broken Nose—Overview: System Behaviors 66
The Worst Game Ever and Martin Luther King Jr. as an Environmentalist—Interdependence 69
Pickup Games and Hungry Ants—Self-Organization 74

Mayan Ball Games and Chimpanzees—Emergence 77
Unfair Goals and Lewis's Lizard—Resilience 80
Zidane and Disappearing Ice—Threshold Crossing 83
Review 88

5. EVALUATING 89

To evaluate sustainability in our systems, and check our progress, there are properties to consider and methods to apply.

Mozambique's Civil War and Portugal's Best World Cup—Overview: Evaluating Sustainability 89
Lampard and Me, Highways and Railroads—Path Dependence 92
Panenka's Gift and New Jersey Dune Grass—Inertia 95
Argentine Defenders and Unsuicidal Lemmings—Carrying Capacities 99
My Missed Penalty and a Stern Review—Counterfactuals 104
Barbosa, Bigode, and the Choice to Eat Dirt—The Five Whys 107
Footprints of the World Cup—Life-Cycle Assessment 112
Review 118

6. CREATING 119

Once we've defined and evaluated our systems, we're ready to create systems that are more sustainable.

Trees in Maputo and the Sacred Monster in Amsterdam—Overview: Creating Sustainable Systems 119
False-Brooding Runs and Wind Turbines—Biomimicry 122
Soccer-Shirt Quilts and the Recycling Distraction—Closing Loops 125
The Goalkeeper Pick Trick and Irish Lumpers—Adaptability 128
"Los Galácticos" and New England Lobstermen—Polycentrism 131
The Fan Who Scored for West Ham and Divestment—Transparency 135
"Take the Piss" and "Let Them Eat Cake"—Fixing Inequality 139
Breakaways, Pass-Backs, and My Repurposed Office—Elegance 142
Review 146

7. THE ENDLESS QUEST 147

There is no magical fix; pursuing sustainability is an endless quest of constant effort at the limits of our abilities, and that's where the fun is.

Maputo, Mozambique, and the World—Overview: The Endless Quest 147
Cruyff and Costa Rican Carbon Neutrality—Visioning 149
Offsides and Refrigerators—Rules 153
Bertha, Dilma, and Marta—Leverage Points 156
The Greatest Leverage Point and Marta Continued—Mindsets 161
Review 164

Glossary 165
Recommended Reading 167
Index 173

Acknowledgments

I'm acknowledging just eleven of the many people who contributed to this book because that's how many players are in a soccer starting lineup. My acknowledgment team plays in a 3–5–2 formation: three defenders, five midfielders, and two forwards (and a goalkeeper). Argentina used the 3–5–2 as they won the 1986 World Cup with the superstar Diego Maradona.

My goalkeeper is Lynda Kong, who created the drawings. As the only players allowed to use their hands, soccer goalkeepers are unique; they even practice off by themselves. Lynda is like the best goalkeepers, working away in isolation and then rising to the occasion when her unique skills are needed.

On defense are Megan Brovan, Grace Greene, and Caroline Hensley, who helped me write. Playing with just three defenders is rare. It only works if you are lucky enough to have players who are talented, confident, and coordinated—and I did!

Among the five midfield spots, the two defensive midfielders anchor the team. They organize, communicate, and deflect credit while selflessly filling in where needed. That describes my editors, Blake Edgar, Merrik Bush-Pirkle, and Kate Hoffman, and everyone else at UC Press (yes, I know that is more than two). The two wing midfielders run up and down the field doing tireless and invisible work. Paulo Coelho, my unpaid agent, plays this role, and so does Tammy Stokoe, my high school English teacher. She convinced me, and countless other teenagers, that writing is fun—but only after earning our respect by attending our soccer games. The attacking midfielder tries to score and make goals, working hardest when there is the potential for tangible rewards. That was my favorite position to play, and it's also what I got to do in writing this book. Rather than acknowledge myself, I'll thank my two main muses: Bruce Springsteen, whose music has serenaded me for thousands of hours of writing, and Bill Simmons, whose best writing proves that serious and funny can go together—and that books full of footnotes can sell.

One of the forwards in the 3–5–2 plays as far up the field as possible, to stretch out the opponent's defense. My brother Rick and his wife, Christine Moskell, played way up the field for this book. They helped generate stories by semi-seriously pondering questions like "How does German soccer explain urban forestry?" And my brother double-checked many of the soccer stories, so any mistakes in those are his fault. The other forward is where the legend Maradona played, and without him, none of Argentina's other pieces would have mattered. Lining up here is my partner, Monica Patterson, with our son, Ezra, in her arms. Monica smoothed my transition from player to person. She changed my utilitarian mindset, helping me appreciate art and beauty as the reason why everything else matters, or maybe vice versa. Without Monica, Marta would not be a hero in this book.

1

Background

On sustainability, what soccer and systems-thinking have to do with it, and how I'm qualified to explain.

MY DELAYED EPIPHANY ABOUT SUSTAINABILITY

Like most of my epiphanies, it's embarrassing I didn't have it sooner. I was mentally processing some of my recent reading as I walked home after a rewarding day of advising graduate students and teaching two of my favorite college courses. On that walk is when it finally sank in for me—the connection between climate change and human rights.[1] Of course, climate refugees who have been robbed of their right to shelter already know this. Some refugees are forced to uproot their entire lives, which is a comparatively good option. Those who lack the financial or physical ability to move remain stuck in the same vulnerable location, except now without shelter.

It turns out there were about thirty-six million climate refugees in 2009 alone,[2] which was the year I finally had my epiphany. Or to put it another way, if you had randomly traded places with someone else in the world that year, you would have been four times more likely to be a climate refugee than a resident of New York City.

Climate refugees face harsh realities. They endure violence against migrant groups as they struggle for their share of overstressed food and

1. Desmond Tutu, who learned a thing or two about human rights while bringing down apartheid, warned in 2014 that "climate change has emerged as the human rights challenge of our time."

2. The United Nations High Commission for Refugees reported that thirty-six million people were displaced by disasters in 2009.

water resources. Such pressures are one reason the U.S. military recognizes climate change as a security threat multiplier. More refugees equals less stability. And we can be certain there will be more climate refugees the less we do about climate change.

My education should have led me to my epiphany sooner. I had read hundreds of thousands of pages about climate change and about human rights. I had used *sustainability,* a term that encompasses both climate change and human rights, in the title of my engineering Ph.D. dissertation. I had spent four years at a liberal arts college proud of its long history of teaching students to "ignore boundaries" and "make connections," presumably between topics such as climate change and human rights.

My work experiences should have led me to the epiphany sooner. That same term *sustainability* was in the title of those college courses I was teaching at the time of my epiphany. I had helped design and build a solar-powered home on the Northern Cheyenne Indian Reservation. I came to the reservation eager to reduce climate-changing emissions from home energy use. I left the reservation hoping that fewer parents would need to choose whether heat or food was the bigger necessity for their children.

My personal life should have led me to the epiphany sooner. I married a human rights advocate, by far the best connection I made in college. My dad, a biologist, planned family outings to teach us about connections in the natural world. My sister, brother, mom, and I got to stay up late for salamander migrations (which are not that impressive) and woodcock mating rituals (which are).

When our family dog had other commitments, I was a bodyguard/witness for my dad as he knocked on cabin doors to ask hunters' permission to study streams running through their property. My dad probably thought it was obvious to me that he studied the algae and phosphorus in the streams because of connections to the water we drink and food we eat. I just assumed he really liked algae and phosphorus.[3]

Even my soccer life should have led me to the epiphany sooner. I had devoted an irrational amount of physical and mental effort to a sport that, as we're about to see, can reveal unexpected connections.

Again, I'm embarrassed that I saw climate change and human rights as isolated problems before that walk home. And I'll never know what,

3. I have since learned that too much phosphorus in a stream can lead to algal blooms, which lead to low dissolved oxygen levels, which lead to the death of aquatic animals. In many cases, the skewed phosphorus levels can be traced to runoff from overfertilized crops and lawns.

specifically, finally shifted my mindset after a lifetime of relevant experiences.[4] But I now recognize the core ideas that brought me to that insight and to others like it. And these ideas seem to also bring epiphanies for the thousands of students I have had the privilege of working with.[5]

So, I think I can help your sustainability epiphanies come sooner than mine. That's why I wrote this book.[6]

WHY SOCCER?

Sócrates Has Our Answer

Soccer is part of life for billions of people. It is both the most popular and the fastest-growing sport on Earth.

Unless you are a citizen of Guam, Brunei, Bhutan, or Mauritania, a soccer team representing your nation competed for one of thirty-two coveted spots in the most recent World Cup, which was held in 2014 in Brazil. And even Bhutan has a team trying to qualify for the next World Cup, in 2018 in Russia.

The professional soccer landscape extends far beyond the World Cup. National teams compete for bragging rights in regional and continental tournaments. Professional club teams play each other in domestic leagues and across national borders. In the Champions League, an annual competition between the best club teams in Europe, even preliminary matches can draw bigger live global audiences than American football's Super Bowl.

Professional soccer gets the media attention, but amateur, informal, and even spontaneous play is the lifeblood of soccer. The Game[7] is played and watched for fun in every corner of the world. It happens on fields of grass, sand, asphalt, cobblestone, and dirt. Players may be tall or short, poor or rich, young or old, and skilled or not. And yet, despite these visible differences, the Game is basically the same.

4. I'm pretty sure the book I was reading at the time of my epiphany was *Common Wealth: Economics for a Crowded Planet* by Jeffrey Sachs. However, I had already been exposed to the main ideas in the book, so I don't think I could have just read *Common Wealth* five years earlier and had the same epiphany.

5. My teaching seems to produce epiphanies for many students, but others just find me disorganized.

6. OK, so I also wrote this book for fun, and to hopefully share some of that fun with you.

7. I will capitalize "Game" when referring to the sport of soccer for a few reasons: for clarity; to allude to the larger meaning of the sport; and because nouns are capitalized in German, and Germany won the most recent World Cup.

Because soccer is so far-reaching and engrained, it affects our lives more than any other sport. The Nobel Prize–winning philosopher Albert Camus gave soccer credit for "all that I know most surely about morality and obligations." The Game has caused[8] and paused[9] wars.

But this is not one of those books (books I love, by the way) about how the Game has some larger meaning. Instead, this book responds to a question from Sócrates: "What if we could one day direct this enthusiasm that we have for football[10] toward positive causes for humanity?"

Now, there was no soccer in ancient Greece—at least as we know it.[11] So our question could not possibly have been asked by Socrates, the ancient Greek philosopher. Instead, the "Socrates" who posed our question was Sócrates Brasileiro Sampaio de Souza Vieira de Oliveira, the legendary Brazilian soccer player.[12]

His Greek namesake would have been proud. The Brazilian Sócrates engages us in the Socratic method because his question leads us to our answer—we'll use soccer stories to align passion for the Game with the quest for sustainability, which is our "positive cause for humanity."

Learning about sustainability through soccer is hopefully more fun—and therefore more memorable—than the analogies to made-up water reservoirs through which I've had to learn. Plus, real-world interdependencies, not just analogies, link the soccer system and the systems we hope to sustain; it's just that these connections are not usually obvious. So,

8. The underlying tension of what is known as the "Soccer War" was migration between El Salvador and Honduras. Still, the war was sparked by rioting during soccer games as the two countries competed to qualify for the 1970 World Cup.

9. The same year as the Soccer War, a three-day ceasefire was declared in Nigeria's civil war to accommodate the visit of soccer legend Pelé with his Brazilian club team, Santos. And during World War I, when German and British troops stopped fighting on Christmas day in 1914, they played soccer.

10. "Football" is what pretty much every country in the world calls what Americans refer to as "soccer." I'll use the two interchangeably and use "American football" when I need to refer to the sport in which massive men in helmets and tight pants smash into each other and shorten their life expectancies.

11. There was no soccer in ancient Greece, but, as we'll cover later, the Greeks did play a game called *episkyros,* which had field linings similar to those in soccer.

12. Sócrates was a leader on Brazil's 1982 World Cup team, a team that played some of the most beautiful soccer ever seen. Beauty was more important to Sócrates than the outcome of the game, which is good because Brazil did not win the World Cup in 1982. Not only was Sócrates a soccer great, he was also a political activist, musician, author, and medical doctor. He was addicted to alcohol and nicotine. In the midst of Brazil's military dictatorship, Sócrates democratized Corinthians, the club he was playing for at the time. He also fathered six children and said, "I like to reproduce." Needless to say, Google him.

when we discover these interdependencies in soccer and sustainability, we sharpen our ability to find them in other seemingly unrelated systems.

Sustainability requires a systems view and so does soccer. It's a holistic sport, in which a slight change in one play can affect what follows in unexpected and dramatic ways.[13] Appreciating the Game requires us to do more than simply reduce it to specific moves or plays. We must also expand our perspective to appreciate the infinite possibilities that result because all of these moves and plays are intricately woven in a web of interdependence.

We can understand other sports through statistics from independent events, such as pitches in baseball or downs in American football. Even free-flowing sports like basketball and ice hockey have fewer players than soccer and, therefore, less variability. These sports also rely far more than soccer does on predetermined plays that are practiced repeatedly before the game and then prescribed by coaches during it. In soccer, no amount of simulation in practice can provide the exact scenario encountered in a game. Every moment is unique.

With that in mind, let's return to answer our slightly modified version of Sócrates's question: How can we use soccer on our quest for sustainability?

First, we can tap into the passion for soccer shared by billions of people.

And second, we can learn from the interdependencies in the Game to help us discover connections between our sustainability obligations (such as human rights and climate change).

THE BLACK PANTHER AND THE SACRED MONSTER

Systems-Thinking for Sustainability

At the most basic level, pursuing sustainability means trying to meet our present needs without ruining the ability of people in the future to meet their own needs. The United Nations calls sustainability "the framework for efforts to achieve a higher quality of life for all people."[14]

13. In Germany, *Der Ball ist rund* ("the ball is round") is an overused expression meaning that anything can happen in soccer.

14. This *sustainability* definition is anthropocentric, emphasizing humans. But of course we are linked with plants, animals, and microorganisms in a network that makes life possible. For instance, we rely on plants' photosynthesis for our food and to convert carbon dioxide into the oxygen we need. Plants, in turn, depend on microorganisms fixing nitrogen at their roots.

Climate change is not the only warning sign on our unsustainable path. We are also flirting dangerously with planetary limits when it comes to species loss, pollution of air and water, and deforestation.

It helps me to think of sustainability as sharing the chance to flourish on Earth with as many people as possible, both now and in the future. I also try to remember that sustainability is an idea like freedom, liberty, or faith: we start with a general meaning, and the best way to refine it more is to try to put these ideas into practice.

This broad idea of sustainability is at the core of our biggest challenges: providing food, shelter, and clean water for all; preventing (and adapting to) climate chaos; ensuring that our consumption does not overwhelm the carrying capacity of our only planet; and protecting rights to happiness, political participation, and a clean environment—regardless of race, gender, economic status, or any other differences. I think you get the idea—sustainability applies everywhere we look.

So let's leave the sustainability theory and applications at that for now—I promise you'll learn more through the stories that follow.

No matter the specific application, systems-thinking is the map for our sustainability quests. Systems-thinking is a shift in perspective from the parts to the whole, from objects to relationships, and from structures to processes. The shift in perspective reveals connections we may otherwise overlook. The shift moves our focus from reducing unsustainability toward creating true sustainability.

Systems-thinking complements the more familiar reductionist approach, in which we take things apart and then study the pieces in more detail. Reductionism underpins most academic learning, for which we split reality into courses, majors, disciplines, and specializations. By narrowly defining perspectives, reductionism makes numerical measurement possible and provides an illusion of certainty. But pure reductionism fails us because even when we know all the parts, and even when we know their arrangement and movements, we still have gaps in our knowledge.

Full understanding doesn't come from simply breaking systems into their smallest pieces. In fact, the most essential properties are often due to the relationships between parts. Our brain and eyes are amazing organs on their own, but without the integration between them you wouldn't be reading this.

In the same way, sustainability challenges like the ones from my epiphany (climate change and human rights) cannot be met with a reductionist approach alone. The systems approach shows us connec-

Reductionism	↔	**Systems-thinking**
parts	↔	*whole*
objects	↔	*relationships*
structures	↔	*processes*
certainty	↔	*insight*
quantities	↔	*qualities*
measuring	↔	*mapping*
disciplinary	↔	*multidisciplinary*

Systems-thinking complements reductionism.

tions between the parts of these complex challenges, and between the challenges themselves.

Let's bring in our first soccer analogy to emphasize how systems-thinking complements reductionism.

The reductionist view is sufficient to see the greatness of Eusébio da Silva Ferreira. Eusébio earned his nickname, "The Black Panther," by combining catlike speed and agility with exceptional ball skills. Over his twenty-two-year career, the Black Panther averaged over a goal per game in a sport in which players who score once in every three games are exceptional. Eusébio scored more than six hundred times for Benfica, a Portuguese professional club team. Playing for Portugal's national team at the 1966 World Cup,[15] Eusébio scored nine goals, more than anyone else at that tournament. Goals are the currency of soccer, so we can simply count the ones the Black Panther scored to measure his influence. The reductionist approach works just fine here.

15. Mozambique was a Portuguese colony in 1966, so players from Mozambique represented Portugal's national team. Even as African countries have gained independence, the best African-born players often represent other nations where they also have citizenship. For example, France won the 1998 World Cup thanks to irreplaceable contributions from Marcel Desailly (born in Ghana) and Patrick Vieira (born in Senegal).

On the other hand, we need the systems perspective to fully appreciate players like Mário Coluna, and to understand how he earned one of the best nicknames ever: "The Sacred Monster." Coluna possessed speed, agility, and skill—like Eusébio. But instead of dominating games by scoring goals,[16] the Sacred Monster made his mark in other ways. He stifled other teams' attacks and created countless scoring opportunities for his teammates, including Eusébio.

Like Eusébio, Coluna began his career in Maputo, the largest city in Mozambique. Coluna also starred for Portugal's national team in 1966 and is a legend at Benfica, the club he led to two consecutive European club championships, the first without Eusébio.[17] Coluna didn't score as many goals as Eusébio. Instead, Coluna disrupted opponents. He made his teammates better—whether in games, in practice, or in the locker room. So, while the Sacred Monster's contributions were less obvious than the Black Panther's, they were just as vital.

Systems-thinking is a way to catch what reductionism can miss: connections, relationships, patterns, processes, and context.

A reductionist approach shows us that the object under the microscope has carbon and oxygen elements. With a systems approach, we see whether these elements are arranged to form coal or a diamond.

A reductionist view tells us about Eusébio, the Black Panther. A systems view reveals the beautiful contributions of Coluna, the Sacred Monster.

Both approaches are needed, of course, but reductionism is engrained and has become intuitive while systems-thinking gets overlooked. Ambiguity about systems-thinking is a big reason for this oversight. So, throughout this book, we'll clarify this powerful idea through stories about Adidas's selfish attempt to improve the 2010 World Cup, the development of Florida's Everglades, and lollipop-sucking soccer managers, among others.

Systems include small ecosystems, like my dad's algae-filled streams, and big ecosystems, like the disappearing Amazon rainforest. These natural systems are all connected to some degree. For example, when we

16. As far as who wins the game, it doesn't matter *who* on the team scores, but being the one who scores is typically better for salaries and endorsements—and for being remembered in (most) books.

17. Benfica missed out on three consecutive European Championships when Coluna was injured by a bad foul from an AC Milan player in the 1963 final. Incredibly, the rules then did not allow any substitutions, even for injuries caused by fouls. So not only did Benfica lose Coluna—they also had to play ten versus eleven for the rest of the game.

clear the Amazon rainforest, the carbon released from the trees into the atmosphere contributes to climate change, with the result that people in Maputo, and coastal cities like it all over the world, face rising seas and damaging storms on top of droughts one year and floods the next.

Natural systems are linked to each other and to the systems humans create: everything from buildings, roads, and machines to social systems like governments, corporations, and communities. As with the natural systems, the behavior of these human-designed systems depends on the interconnected elements, flows, stocks, and purposes that we will learn more about in chapter 2.

Entire programs of study focus on selected natural or designed systems.[18] A growing academic discipline is even devoted entirely to systems in general. And researchers spend careers seeking cross-cutting systems principles. This effort is justified because systems-thinking is at the core of some of the biggest scientific contributions in recent history. Among the groundbreaking advances in Albert Einstein's theory of relativity, for example, is the notion that space and time are not separate; the two should be considered together and in relation to each other. Similarly, in medicine, the placebo effect is when fake treatments lead to real improvements, just because patients expect to feel better. Like space and time, the body and mind must be considered together.

If we covered every last detail about systems in this book, I'd get bored, and your path to epiphanies would be even longer than mine. Fortunately, the most valuable use of systems-thinking is also the most straightforward: adding it to our everyday thought processes as we pursue sustainability.

So let's move on, because sustainability and systems are just ideas, and we need examples, applications, and stories to bring these ideas to life.

WHY ME?

An Autobiography in Less Than Twelve Hundred Words[19]

Now that you know it took me three decades to appreciate the connections between climate change and human rights, you're probably wondering

18. Even in these systems-themed courses and majors, too much specialization can lead to the ironic situation where an overly reductionist approach is applied in an ill-fated attempt to study a system.

19. Not counting footnotes.

why you should listen to anything I have to say about sustainability, or systems-thinking, or even soccer for that matter. It's a valid concern, and one I think is best answered with a very brief autobiography.

Until I was six, the upstate New York town where I grew up didn't even have a soccer league. So my dad and his friend started one. They also brought in expert coaches to run summer camps. I knew they were experts because they had accents (which instantly boosts one's soccer credentials in the United States). Despite the camp coaches' accents, my teammates and I learned far more about the Game from my dad.

I only played locally until an all-star team from our league got clobbered by a team from a nearby town. Then my dad and his friend started a travel team for my friends and me to play against better competition. We got pretty good, mostly stayed out of trouble, and learned life lessons about collaboration, persistence, and never trusting referees.

My rural upbringing combined with my driven (and slightly obsessive) personality meant that I spent lots of time practicing soccer alone.[20] I launched shot after shot from our gravel driveway against a homemade wooden backstop attached to the storage shed. For the six months of the year when the ground was frozen or covered with snow, I played in the loft of our repurposed dairy barn.[21] I worked on my skills by dribbling in figure eights to songs by Bob Dylan and Green Day. Then I went outside and ran as fast as I could through the snow.

When I was fifteen, I was invited to join a select travel team by a mature-beyond-his-years future friend whom I saw only as a competitor

20. I worked harder than anyone I knew except my younger sister—who, as in everything else, kept me from resting on my soccer laurels. I started for my high school team as a ninth-grader; she did it in eighth grade. I made the statewide select team; she made the team for the entire Northeast region. I played college soccer at Lafayette College, where an ambitious goal was to be a top-twenty-five team in the country; she played at the University of Maryland, where an ambitious goal was to win a national championship. It hasn't stopped. Some people call me "Dr. Klotz," because I have a Ph.D.; my sister is the real doctor, the kind they ask for on airplanes.

21. When my "little" brother Ricky was nine, I convinced him to play goalie for me in the freezing-cold barn loft. I gave him a thin mat to dive on, which protected him from landing directly on the wooden floor, but the barn was so small that my shots were from closer than ten yards, which meant he could barely get his arms up to protect himself, let alone intentionally stop shots. Five years younger and roughly half my size, Ricky did have one advantage when we switched to playing one-on-one: he could run closer to the walls, which angled downward so that there was more space the closer to the ground you got.

at the time. My parents,[22] who could ride their bikes to work before my sister, brother, and I scrambled their lives, put hundreds of thousands of miles on their Dodge Caravan so I could compete against the best players around. A former star college player coached us for free. We won a national indoor championship, partly because we had talented players and a devoted and clever coach, and partly because the six months of winter meant we practiced indoors a lot more than teams from less dreary climates.

I arrived at Lafayette College the season after they made it to the final sixteen at the top level of college soccer in the United States. With the same coach—and better players, or so we thought—we lost three times as many games as we won in each of my first two seasons. But then some of the "better" players left, the coach changed, and so did our mindsets. We had one of the biggest turnarounds ever in college soccer. In each of my final two seasons, we won our league and qualified for the national championship tournament.

Throughout my amateur career, my job was to score goals, make passes that led to goals, and occasionally play some defense, which mostly meant (illegally) holding the opponents' tallest player on corner kicks. I had more than my fair share of fantastic teammates who freed me to play my role, and I played it well enough to get a chance to keep playing after college.

To participate in professional preseason soccer, I commuted in my early graduation gift, the now old-enough-to-be-cool Dodge Caravan. I missed college classes and parties. Eventually, to the chagrin of my grandmother,[23] I even skipped the graduation ceremony.

After signing my first contract, I cried tears of joy on the Pennsylvania Turnpike somewhere near Altoona as Dylan's "Like a Rolling Stone" brought flashbacks to long and cold hours practicing in the barn. Now that I was officially a professional soccer player, I had achieved a lifelong goal—which I had conceived when the United States had no viable professional soccer league.

I played two seasons, making about $2,000 a month, enough to not have another job (but not enough to indiscriminately buy cheese at the

22. My mom was the only person as vital to our soccer careers as my dad was. You'll hear more about her, in particular when we discuss T-shirt quilts and distracting recycling.

23. I later found out that my grandmother had missed out on her college graduation because the United States entered World War II and she had to immediately serve as a replacement for schoolteachers who were called into action. Her reason for missing graduation is better than mine.

supermarket). My entire compensation as a professional was a fraction of what my parents had spent on my amateur career. I scored a few goals,[24] saw the country from bus and airplane windows, and nutmegged some of my heroes.[25] But I also spent more time on the bench than on the field.

I retired from soccer, but I kept my goal-oriented personality—which is a good thing, because I had about $500 in my bank account, and everything I owned could fit in my used four-door Nissan[26] (if I deflated my air mattress and put my futon frame on the bike rack).

I expected to miss soccer after retiring, but by that time I was hopelessly in love with my then girlfriend, now wife,[27] Monica,[28] and was looking forward to seeing her more. Fortunately, playing college soccer at Lafayette had required that I also earn a degree (in engineering), which made it pretty easy to find a job working on school construction projects in New Jersey. Monica and I lived at the beach, and then in Bruce Springsteen's hometown (Freehold). We could pay for whatever we wanted (she has simple tastes), I golfed a few times a week . . . and I was bored out of my mind.

24. The Portuguese legend Eusébio and I both finished our careers with the exact same number of appearances (five) and goals (one) in the U.S. professional indoor league. Eusébio played for the Buffalo Stallions in 1980–81, at the very end of his career and after having several knee surgeries. I played for the Harrisburg Heat in 2000–01 after recovering from a leg fracture. Unfortunately for me, these parallels do not extend to our outdoor careers. No one knows exactly how many goals Eusébio scored, just that it's over six hundred. No one cares how many I scored (three).

25. A "nutmeg" in soccer is when you dribble the ball between the defender's legs. It's a functional move that also embarrasses the defender, which typically earns a joking cheer from the crowd. I nutmegged Frankie Hejduk on May 27, 2001, in a game in Pittsburgh, Pennsylvania. Hejduk played for the United States at the 1998 and 2002 World Cups. On New Year's Day of that same year, I nutmegged Bob Bradley in Princeton, New Jersey. Bradley managed the U.S. team at the 2010 World Cup. Both Hejduk and Bradley had far more accomplished careers than I, and I'm sure neither would consider it a highlight of their career had they nutmegged me.

26. When I bought the Nissan, my brother got the Dodge Caravan, which now shook violently when driven at highway speeds, perhaps because the front bumper was now attached with wire.

27. Among the attendees at our wedding was a full side of eleven teammates, plus a coach, a utility player, and an extra goalkeeper (for adaptability, which will make sense later if it doesn't already).

28. Soccer was the only reason I had any luck with girls as an always shy, sometimes cruel, and often acne-plagued adolescent. When Monica and I met, she didn't even know I played.

I still had great times with Monica, my family,[29] and my friends, but I no longer had a guiding focus to provide meaning to my life. And yes, I do recognize how silly it is to think that winning a soccer game is meaningful and building schools is not. At least you know I'm being honest.

My career transition from construction manager to professor was part of this search for meaning, and it has mostly worked. Nothing I do now is more rewarding than spending time with bright and motivated students. Research challenges me to always learn new things. And just when I start to think the mental struggle is futile, research is rewarding in an entirely different way, with unexpected breakthroughs. Admittedly, the goal I scored against our college rivals in the league championship game still gives me more pride than any of my published research papers, but the gap is closing.

The fundamental concept behind all of my teaching and research is systems-thinking. Sustainability is the application.

So now you know that there were better soccer players than me. And, fortunately for humanity, I'm not the only one devoting my career to sustainability. But when it comes to soccer *and* sustainability, I'm your guy.

HOW TO USE THIS BOOK

I hope this book both rewards and absorbs you. To maintain flow in the text, I use footnotes for bonus information that I think is insightful—or funny. If you disagree after reading a few footnotes, you can skip them and breeze through the main text.

Footnotes also include seminal supporting references. Of course, I haven't listed every last reference for the wide range of topics we'll cover, especially in cases when a Google keyword search will quickly give you evidence and insight. Google can confirm the soccer stories too, and a Youtube search will usually turn up supporting video.

29. When I finished playing, Rick was all of a sudden bigger than me. I cheered for his high school and college teams, sitting with my parents (in the offensive end) and getting more nervous than I ever had while playing. On his summer break from college, Rick occupied the living room of the one-bedroom apartment Monica and I shared. Rick and I spent hours practicing heading, and I'll never forget him scoring on a header in one of the first games of the next season.

But we can't leave everything to search engines. So, in addition to the seminal references in the footnotes, I've shared my favorite additional resources in a "Recommended Reading" section at the end. I hope this approach balances your desire to learn more with the reality that the best learning is self directed.

To get the most from this book, you can stop and reflect after reading each section. Look at the drawing and try to figure out how it represents the stories and ideas you just read about. Then find your own example of the main sustainability idea from that section. If you're a soccer fan, think of a soccer example too. And then talk to someone about what you came up with.[30]

30. If Twitter still exists when you are reading this, I'd love to see your examples. My handle is @leidyklotz. Use the hashtag #sustainabilitythroughsoccer.

2

Parts

Systems-thinking is the guide for our sustainability quest, and systems are made up of some basic parts: elements, flows, stocks, feedback loops, and purposes.

MAPUTO AND THE SACRED MONSTER

Overview: System Parts

Systems-thinking guides our quest for sustainability, and it reveals the greatness of Mário Coluna, the overlooked Sacred Monster. In order to appreciate how systems-thinking can help us, we need a shared understanding of systems.

Systems come in all different sizes: from organisms to cities to societies, from soccer players to games to tournaments. Systems are connected to other systems, are made up of smaller systems, and are part of larger ones.

Systems of all types, whether cities or soccer teams, have some basic parts: elements, flows, stocks, and feedback loops. When the connected behavior of these parts leads to a purpose, we consider it a system.

A system's building blocks are *elements*. The people, buildings, and fiery red acacia trees in Maputo, the capital of Mozambique,[1] are all elements of that city system. Coluna was a boxing and high-jump champion in Maputo before applying his work ethic, intelligence, and athleticism to soccer. Those traits would become game-changing elements in future soccer systems.

1. Mozambique is in Southern Africa, just across the Indian Ocean from Madagascar. Mozambique is closer to Antarctica than it is to Portugal, which controlled the region for centuries.

Elements are vital parts of any system. Yet whether our goal is to fully appreciate the Sacred Monster or to help make a more sustainable Maputo, we also need to know the arrangement of these elements and how they interact.

One way elements interact is through movements called *flows*. In his role as a midfielder, Coluna spent less time near the opponent's goal than forwards like Eusébio. Because of the different flows that go along with different positions in soccer, the Sacred Monster had fewer scoring chances than the Black Panther. The arrangement made their teams better because once the Sacred Monster was deployed in the midfield, the two star players could cover the entire field with their flows.

Flows and elements build up in a system and become *stocks,* which reveal what has already happened in the system. In soccer, goals are a stock that tells us how the game is going. Similarly, population is a stock that reflects whether a city is growing or contracting. Maputo's population grew after a railroad connected it to Pretoria, South Africa, in 1895. Population decreased (very slightly) when promising soccer players like Coluna and Eusébio made the thirty-four-hour boat journey to Portugal. When Mozambique gained independence in 1975,[2] Maputo's stock of people shrank as hundreds of thousands of Portuguese left.

Elements, stocks, and flows may be new terms for you, but they are not new ideas. In fact, a reductionist approach can be enough to appreciate these parts of a system. The problem is that when they are viewed in isolation, elements, stocks, and flows show an incomplete picture of the system. Simply knowing the population of Maputo tells us nothing about the quality of life there.

To seek sustainability and to appreciate the Sacred Monster, we have to consider the learning that changes elements, flows, and stocks—and therefore determines system behavior. Systems are dynamic: information from the past shapes future actions; effects are causes, and causes are effects.

Systems learn and evolve through *feedback loops,* which are connections between a stock, actions dependent on the stock, and a flow or element that is changing the stock.

The Sacred Monster reinforced a feedback loop that made the Black Panther's career possible. Coluna moved from Mozambique to Portugal in 1954, and by 1961 he had led Benfica to four Portuguese league cham-

2. Catalysts for Mozambique's independence were a decade-long war for self-rule in Mozambique and an anticolonial government coming to power in Portugal.

pionships and their first European Cup title. With his performance, Coluna showed that players from Mozambique could make Benfica better. This opened the door[3] for the younger Eusébio, whose pairing with Coluna led to another European Cup title and seven more Portuguese league championships. The stock (Benfica players from Mozambique) was connected to actions dependent on that stock (Benfica's performance) and to a flow to change that stock (players from Mozambique getting opportunities with Benfica). Reinforcing feedback loops like this one lead to an accelerating rate of change.

Feedback loops can also slow the rate of change. A balancing feedback loop is the underlying rationale for those who claim that privatizing supply will increase access to clean water in developing cities like Maputo. The claim raises some serious questions,[4] but the basic idea is that limited access to clean water (the stock) will inflate the prices (the action depending on that stock), which will incentivize companies to provide more water (a flow to change the stock), which will increase access to clean water (the stock).

The behaviors of elements, flows, stocks, and feedback loops all depend on a system's *purpose,* which is what the system is trying to accomplish. Without any immediate changes to its physical layout or population, Maputo was transformed into a far more sustainable system after a fifteen-year civil war ended in 1992 and the purposes of the warring groups switched from military conflict to political competition.

New purposes change soccer careers, too. Coluna was a prolific goalscorer as a young player in Mozambique; he scored seven times in one match against a visiting South African club.[5] Even during his first season in Portugal, Coluna played in a mostly attacking role and led Benfica to the league title. Coluna scored fourteen times in twenty-six games—less than the Black Panther's goal-per-game rate, but better than just about anyone else's. It wasn't until his second season in Portugal that Coluna took on the creative and disruptive midfield role that would define the rest of his career. The new purpose cut down Coluna's goal scoring, but it gave birth to the Sacred Monster.

3. Both Coluna and Eusébio began their Lisbon life in Benfica's Lar do Jogador, a residence where the club housed players who were new to the city.

4. Questions to ask when considering privatizing the supply of water: What about people who have no money to pay for the water, which is a basic need? And how can you prevent an unfair setup whereby the public pays and the private profits?

5. Coluna didn't score when the same two teams played in South Africa, but only because apartheid laws prevented him from stepping onto the field.

PEOPLE AND PLAYERS

System Elements

My Brazilian friend Ze and I wanted to get some fresh air and sunshine after our morning practice with the Harrisburg Heat professional indoor team. So, Ze grabbed a tennis ball and we headed outside to the parking lot by our apartment. I thought we'd play catch. But once we were outside, Ze began to juggle that tennis ball—with his feet.

Ze's juggling blew my mind. I had spent most of my ample free time as a nine-year-old learning how to do the same thing with a soccer ball. Juggling a tennis ball is infinitely more challenging. That day, I never got more than eight juggles in a row without the tennis ball hitting the pavement. Ze kept the ball in the air as long as he wanted. When I expressed amazement at his ability, Ze asked me in his broken English (which was far less broken than my Portuguese), "Why you think Brazil is the best in the world?"

I'll never forget Ze's rhetorical question.[6] In just ten words, he politely conveyed that he was better than me at soccer, explained why Brazil is historically the best national team (because they have the best players), and boiled the Game down to its essence—those who play it.

Ze has had a long and successful[7] professional career in the United States, especially compared with mine. I was probably somewhere around the hundredth best player born in this country in 1978. Ze was also born in 1978, and I bet there were closer to a thousand Brazilians born that year who were at least as good as him.[8] Again, "Why you think Brazil is the best in the world?"

In soccer, players are the most obvious and influential system elements. Sure, other elements of the system matter, from managers to referees to the ball (as we'll see in the next story). But players are the basic elements of the Game; other parts of the soccer system change the outcome only when the players are roughly equal. A team of world all-stars—even if they speak different languages, have never practiced together, and all play the same position—will easily beat a team of local

6. Another Ze quote I will never forget is when he said of patrons at a local shopping mall, with absolutely no malice intended, "I think some of these people do not have mirrors."

7. Ze's full name is José Roberto Gomes Santana, which is also the title of his Wikipedia page, and having a Wikipedia page is one indicator of having a successful professional soccer career.

8. Argentina and Brazil "export" more professional-caliber soccer players than any other country.

fourteen-year-olds. It doesn't matter who is coaching or how many fans are cheering for the fourteen-year-olds.

Elements are the building blocks for the system and visibly shape system behavior.

On our quest for sustainability, people are often the most influential system elements, just like the players in the soccer system. Consider your favorite city.[9] Sure, buildings and roads are visible elements that are vital to the appearance and function of the city. But sustainability also depends on the human element. It depends on people's actions in those buildings and on those roads. How much do we drive? How much energy do we use to heat and cool our homes and businesses?

The buildings and roads shape our actions, of course. But given that people designed and built them in the first place, the fact that this built environment shapes our actions is even more evidence of our sway in the systems we are part of.

One last point about elements: those that appear similar can behave differently. Ze and I both were soccer players. We were the same age and played the same position. But Ze's juggling skill was one of the many actions that made him a more useful soccer player than me. In the city system, actions such as voting or abstaining, taking or sharing, and even cheering or booing[10] all vary by individual, and these actions change behaviors of the larger system.

So, elements are the basic building blocks of systems. But we can't stop there, because the same basic elements are in systems that behave differently. People all over the world are almost identical biologically.[11] Most of our differences come from the systems we are part of. Or put another way: Ze juggled that tennis ball with his feet not because of some innate skill he was born with, but because the skill was nurtured in Brazil's system. "Why you think Brazil is the best in the world?"

Humans and our actions are vital elements in many systems, from cities to soccer games. Think of the behavior of people in your favorite city,

9. Philadelphia is my favorite city, partly because that's where my wife is from and partly because they have the best crusty bread.

10. In Philadelphia, fans of the Eagles (an American football team) booed Santa Claus and threw snowballs at him. The Philadelphia sports fans I know are actually proud that it happened in their city. They will argue that snowballs were thrown not because he was Santa, but because he was a substandard fill-in posing as Santa.

11. Racial differences, for example, are thought to be determined by only about one-tenth of 1 percent of our DNA.

or consider my Brazilian roommate juggling tennis balls (with his feet) in a Pennsylvania parking lot.

FLOATING JABULANIS AND THE RIVER OF GRASS

Overlooked System Elements

As the 2010 World Cup in South Africa kicked off, Argentina's Lionel Messi was the best player in the world[12] and a paid Adidas spokesperson. Not only was the company paying millions for the association with Messi's brilliance, but Adidas had also bought the right to provide the official game ball for that World Cup. They came up with the "Jabulani,"

12. To name the best player each year (and to generate publicity and advertising revenue), FIFA (see footnote 17) established the Ballon d'Or award. Messi won the award in 2010, and again in 2011 and 2012.

a ball meant to fly faster and farther than any that had come before. Yet, despite the good intentions of Adidas and despite having millions of reasons to keep his opinion of the ball to himself, Messi spoke out against his sponsor's creation, saying that "the ball is difficult for the goalkeepers and also for us. We cannot really get used to it."

Messi's take on the Jabulani is generous compared with what was said by those not on Adidas's payroll. Brazilian striker Robinho speculated that "the guy who designed this ball never played football." England's Italian manager, Fabio Capello, called it "the worst ball I have seen in my life."

The Jabulani was so bad, it even got a mention in *Soccer in Sun and Shadow,* the poetic history of soccer by Uruguayan author Eduardo Galeano,[13] who wrote, "'Jabulani' was the name of the soapy, half-crazed tournament ball that eluded hands and flouted feet."

The results on the field confirmed the complaints of the players and poets. Just twenty-five goals were scored in the first round of sixteen games in the 2010 World Cup—far fewer than the thirty-nine goals scored in the first round in 2006, or the forty-nine scored in the 2014 event.

If World Cup goals remain forever etched in your memory as they are in mine, you will recall that in 2010, a few long-distance goals did seem to get an assist from the Jabulani's unpredictable flight.[14] But those goals were the exception; mostly fans got to watch the planet's best players launch shots over the goal and sail passes beyond intended targets.[15]

13. Galeano's *Soccer in Sun and Shadow* is a must-read for anyone who loves stimulating thought, historical soccer, or extraordinary writing. This isn't the last time I'll bring up the book.

14. Diego Forlán (Uruguay), Giovanni van Bronckhorst (Netherlands), and Keisuke Honda (Japan) did eventually adapt to the Jabulani enough (or just got lucky) to score long-range goals in that 2010 World Cup. And thank you, Clint Dempsey and, especially, Robert Green (and the less-than-perfect fields in South Africa) for the long-distance goal that tied the United States' game against England.

15. For those of you interested in the physics, here is my take on the issue. The air moving around the ball is either laminar (smooth) or turbulent (rough). Laminar air movement creates predictable flight that is either straight or a constant curve, like a David Beckham free kick. Turbulent movement creates unpredictable flight, like a knuckling free kick from Cristiano Ronaldo. The air movement around the ball is more likely to be turbulent (1) the faster the ball is flying and (2) the smoother the ball is. So, as Adidas tries to make balls go faster, in part by making them smoother, long shots are more likely to generate turbulent air movement and therefore go out of control. Of course, players can still practice to figure out speeds at which the ball will stay in the steady laminar flow range. Ronaldo honed his technique to take advantage of the new ball's flight. And Italy's Andrea Pirlo—who for the first sixty years (maybe an exaggeration) of his career hit free

Eventually, the uncontrollable ball dissuaded players from taking long-distance shots, which helped defenses even more because now they could devote extra attention to guarding players close to the goal.

Players may be the most obvious element of the soccer system,[16] but the Jabulani reminds us not to ignore other elements, which can also change the outcome of games.

Changes to the ball element of the soccer system distorted the valued skills of player elements of the system. When the ball is uncontrollable on long-distance shots and passes, players who distinguish themselves with their long-distance shooting and passing skills suddenly lose a hard-earned advantage, thanks to the whims of a few ball designers and the money Adidas pays FIFA[17] for the right to provide balls for the World Cup.[18]

You may have guessed from my frustration with the Jabulani that long-distance shooting and passing was my main skill, honed through years of practicing with walls instead of people (which you'll read more about when we get to feedback loops). Because I retired before having to play with the Jabulani, I can't blame it for any of my many misses. Still, I'll never forgive Adidas for robbing soccer fans everywhere of the long-range, earth-shaking goals that were missing from the 2010 World Cup.

We tend to overlook, or even intentionally manipulate, certain system elements—until our negligence is revealed through problems in a more noticeable part of the system.

Unfortunately, this tendency is not limited to floating Jabulanis; overlooked elements are a cause of many of our sustainability challenges.

---

kicks in the laminar range—showed that he has adapted to the new turbulent possibilities with his shot off the crossbar against England in the 2014 World Cup.

16. Players are the most visible element, assuming that we draw our system boundaries around the playing field. If we extend the boundaries to include the stands and the fans in them, sometimes even the players are overshadowed by the impossibly beautiful people cameramen "randomly" find in the colors of the competing teams.

17. Fédération Internationale de Football Association (FIFA) is a "nonprofit" organization that has been compared to an organized crime family, first by the investigative reporter Andrew Jennings and then by the U.S. Department of Justice. According to their own financial report, FIFA's thirty-eight "key personnel" received $33,500,000 in compensation in 2012. This doesn't include bribes, which are not included in the financial report, so you'll have to do a Google search and develop your own estimates.

18. Presumably, Adidas hoped their Jabulani would lead to more goals and more entertaining soccer. I suspect this hope mattered less to them than producing a ball that was different from the norm, thereby suggesting to young soccer players around the world that, if they wanted to stay current, they had to have the new ball, which cost $199.

Consider the problems we are now trying to fix in the Everglades, a region of tropical wetlands in the southern half of Florida. The Everglades are part of a watershed system that begins near the city of Orlando with the Kissimmee River. The river empties into Lake Okeechobee, which, during the wet season, discharges water in a slow-moving grassy river. This "River of Grass" is more than 160 kilometers long and more than ten times wider than the widest part of the Nile River. The ecosystem provides all kinds of services that benefit humans; in particular, the grasses naturally sequester climate-changing CO_2 emissions better than anything we humans have been able to engineer.[19]

The Everglades used to be a huge web of marshes and prairies roughly the size of Iceland, just a lot warmer and with more alligators. With the Industrial Revolution in the 1800s, humans gained the machinery to quickly and cheaply move big piles of dirt. Since then, in mostly well-meaning attempts to help the larger south Florida area flourish economically, humans have relentlessly engineered elements of the Everglades, manipulating the ecosystem to suit human development.

Today, about half of the original Everglades are developed. Swamp elements in the Everglades system were drained to grow food crops, mostly sugarcane. Everglade swamps were drained to make homesites, where some people live and where others own redundant vacation estates that they rarely use and that serve mostly as monuments of wealth and status (and impulsiveness). Swamp elements were manipulated to divert water for the swelling population in the Miami metropolitan area and to control floods in that region, which is ironic because the untouched ecosystem was doing a pretty good job with flood control before we developed it in the first place.

Just as the Jabulani brought less entertaining soccer, development in the swamp elements of the Everglades system brought lower quality of life to south Florida cities that have grown beyond their capacities to sustain themselves. In a region that gets plenty of rainfall, people face declining water quality because the lost wetlands no longer filter polluted water. Because homes supplied with clean drinking water cost

19. Sequestration means capturing CO_2, much of which is produced by burning fossil fuels, and storing it safely away from the atmosphere. Cost-effective methods to capture and permanently store these CO_2 emissions could help us avoid the worst effects of climate change. Such methods don't yet exist, but to try to find them, we can learn from plants like the grasses in the Everglades, which pull CO_2 out of the atmosphere and store it until they decompose (or are burned).

more than homes supplied with dirty water, economically poor people are disproportionately affected.

It's not just those without clean water feeling the effects. People all over the United States are literally paying the price of trying to reverse these trends. The Comprehensive Everglades Restoration Plan was passed into law with an estimated price tag of 7.8 billion taxpayer dollars.

That seems like a lot of money, but in reality, ecosystem services like those provided by the Everglades are priceless—remember, we can't live without an environment that supports us. Ecosystems provide us with food and fresh water. They provide us with shelter: wood for our homes and fiber for our clothes. Beyond those direct services, ecosystems stabilize the climate and disease-transmission patterns that our settlements are built around. Ecosystems enable agricultural productivity through nutrient cycling and pollination. And ecosystems provide cultural services, as you know if you are fortunate enough to have taken a walk in the woods or spent a day at the beach. Precisely translating these priceless services into an economic value is impossible—nature can't be converted into money. But, without estimates, these services will continue to be overlooked by markets. So, while estimates vary, it is conservative to consider that worldwide values of ecosystem services are in the tens of trillions of dollars[20]—per year.

I realize that finding problems and assigning blame is much easier after the fact. Still, it seems to me that systems-thinking might have prevented the Jabulani and Everglades fiascoes by simply reminding us that overlooked elements, whether a ball or a seemingly isolated swamp, can be vital parts of a larger system.

For our soccer example, that reminder might have led to additional testing of the Jabulani. Adidas hired scientists to study the ball, but players didn't use it until a few weeks before the 2010 World Cup.[21] By then, it was too late. Adidas retired the Jabulani shortly after that tournament.

Studying how a ball will behave is difficult enough, so we can appreciate the challenge of anticipating development effects in a vast and intricate system like the Everglades. But it's not hopeless. Designers working to restore the ecosystem apply common sense and mathematical models

20. To get a sense of how much a trillion dollars is, I just think of my annual salary, and then add a trillion dollars.

21. Players in Germany's professional league, for which Adidas is a sponsor, got to use the ball in the season leading up to the World Cup.

to consider regional impacts on water quality and quality of life. And as these designs become real, many of the people paying the bill to restore the River of Grass are descendants of those who paid to change it in the first place.

We take for granted the less obvious elements of systems—until they get screwed up. Examples are Adidas's selfish attempt to add excitement to the 2010 World Cup as well as the development (and now undevelopment) of swamps in Florida's Everglades.

"REMEMBER ISTANBUL" AND THE CROSS BRONX EXPRESSWAY

Physical Flows

Every year since 1955, the best club teams from all over Europe have played each other until only one victor remains. The tournament, formerly known as the European Cup, is now called the Champions League.[22]

Champions League games routinely draw more viewers than the Super Bowl, and these viewers enjoy soccer played at a level that is even higher than in the World Cup. The national teams that compete in the latter include only citizens of one country. National teams play together for only a handful of games each year, which makes it almost impossible

22. Calling it the European Cup is more appropriate—calling it the Champions League without including teams from other continents, in particular South America, is misleading.

to form a cohesive team. By comparison, the richest club teams bring together the best players from countries all over the world, and these players have ample time to gel, typically playing more than fifty games together annually.

One of the most famous Champions League games is the 2005 final, played at the Atatürk Stadium in Istanbul, Turkey. The English club Liverpool FC fell behind, 1–0, after conceding a goal to Italy's AC Milan in the very first minute of the game, the earliest goal ever scored in the history of the European Cup final. Then things got even worse for Liverpool.

Midway through the first half, Liverpool's Harry Kewell, an attacking player who was chosen to start the game despite a recent injury, got reinjured and had to be substituted.[23] Only three substitutes are allowed for the entire game, so Liverpool was put at an even bigger disadvantage by being forced to use one of their substitutes so early in the game.

Following Kewell's injury, AC Milan scored twice more, and at halftime Liverpool found themselves behind 3–0, a nearly insurmountable lead in soccer. Even worse for Liverpool, AC Milan had not allowed a single goal in any of their previous six Champions League games! "Remember Istanbul" appeared destined to become a mocking chant for Liverpool's rivals.

Everyone assumed the game had been decided, but what happened in the second half shows why we must consider physical flows in systems. Liverpool's only halftime player change was replacing the Irishman Steve Finnan with the German Dietmar Hamann. All the other players stayed the same, as did the ball, referee, fans, field, and other system elements.

Liverpool's decisive halftime change was not in personnel, but in switching from four defenders to three. The extra player moved into the midfield to provide support for Liverpool's star, Steven Gerrard, who was then free to attack more. Relieving Gerrard of defensive responsibilities was a calculated gamble that made sense for a team trailing 3–0.

The change worked better than even the most optimistic Liverpool fan could have hoped. After only fifteen minutes of play in the second half, Liverpool had scored three goals to tie the game. With his new attacking freedom, Gerrard had scored on a header and set up another goal by drawing a foul that earned a penalty kick for Liverpool. The

23. My brother believes that Kewell's reinjury and subsequent substitution actually helped Liverpool.

game stayed tied and went to a deciding penalty-kick shootout—which, as you have guessed, Liverpool won.[24]

Now, Liverpool fans proudly belt out the "Remember Istanbul" chant every chance they get.

Admittedly, formation change was not the only reason for Liverpool's comeback. At halftime, Liverpool's players stayed calm and focused in the locker room. Their supporters in the Atatürk Stadium stands kept the faith, confidently[25] chanting "We're gonna win 4–3." On the field in the second half, it was Gerrard and his teammates' skill and spirit that made the tactical change work. And, of course, luck[26] plays a part any time three goals are scored in fifteen minutes, especially when they are scored against one of the best teams in the world playing in the biggest game of the year.

Nevertheless, Liverpool would most certainly have lost the game without the modified post-halftime formation, which changed the physical flows of the system. Such flows are called "tactics" in soccer. Tactics determine which parts of the field, and which opponents, each player is responsible for defending and attacking. Tactics influence where and how players pass the ball, where they run to receive passes, and how fast they move to get there. Tactics can explain why the team with the better players isn't necessarily the better team.

Physical flows are movement over time. These flows modify the behavior of the elements—and therefore of the system.

Physical flows don't just change soccer games; they can shape sustainability outcomes, too. For evidence, we need look no further than the physical infrastructure where we live.

A towering (relatively) six-story apartment complex was recently built right in the heart of our blink-and-you-miss-it downtown in Clemson, South Carolina. The complex is directly across the street from Clemson University and less than a kilometer from where I live. Some residents complain that the new complex ruins the historical feel of downtown.[27] Some make dire predictions that the complex will mean more college

24. Penalty kicks usually don't go well for English teams and do for Italian ones. This game was an exception.

25. My brother insists the "We're gonna win 4–3" chant can only be explained as alcohol-fueled sarcasm.

26. In particular, Liverpool were fortunate to be awarded a penalty kick for the "foul" on Gerrard.

27. I'm skeptical, but maybe tourists *are* flocking to see national sandwich chains Subway, Jimmy John's, Pita Pit, and Firehouse—all within one block of each other.

students milling around, which apparently is not expected in a town of thirteen thousand people that exists because it is next to a university of twenty thousand. Honest residents admit that what they are really concerned about is that more people living downtown will mean more traffic downtown. But if the complex has led to more traffic, I haven't noticed.

Professional planners know that more dense housing does not necessarily increase traffic. When people live closer together, grocery stores, restaurants, and other services are more likely to pop up within walking distance of the people, who then don't need to drive as much. In our local case, Clemson students who live outside of town drive to class—right through the middle of downtown. Students living in the downtown apartment complex walk to class, not necessarily in a conscious effort to avoid spewing climate-changing emissions, but because walking is faster than driving.

Planning, or lack thereof, for shelter and mobility also guides flows of people in places with more than thirteen thousand residents. The Cross Bronx Expressway was built in the mid-1900s to make it easier for automobiles to move through that section of New York City, which is just north of Manhattan. Perhaps the expressway temporarily eased physical flows for commuters lucky enough to own an automobile, but certainly the expressway ruined physical flow for those living in the established neighborhoods that were divided to make way for the new road.

After the expressway was completed, the surrounding South Bronx deteriorated from tight-knit communities to decaying examples of "the other side of the tracks." Unfortunately, the Cross Bronx Expressway was not an isolated case in the United States. All too often, highways literally divided thriving working-class neighborhoods in a futile attempt to make it more convenient for those who wanted to be able to work in the city and live in the country.

In this case, Automobiles create *externalities,* which is when the people getting the benefits are not the same people who are paying the costs, and vice versa. The distortion in automobile costs and benefits goes beyond the expense of building roads.[28] There are over a billion cars in the world, which means there are at least six billion people who do not own a car but still have to breathe polluted air and adapt to climate change.

28. Roads used to be for people; it is only recently that automobiles have become the assumed primary user. Part of the appeal of cities like Florence, Italy, is that they have been able to return people—whether walking, biking, or driving—to their rightful place as primary road user.

One way to avoid externalities like these is to follow the "polluter pays" principle, which is a sort of golden rule that says we all should clean up after ourselves. Similarly, carbon taxes attempt to internalize externalities by charging a certain amount per ton of emissions, based on the social costs of the emissions.[29]

Even if we are able to account for externalities, there is an even more basic problem with simply building more or wider highways: adding roads doesn't relieve traffic for long. You already know this if you've ever traveled the Cross Bronx Expressway during rush hour, which lasts from about five in the morning to eleven at night. In fact, overcrowded roads that were initially built in an attempt to ease traffic are so common that the phenomenon is considered the "fundamental law of road congestion."[30] More convenient driving leads to more drivers, which leads to more congestion—a feedback loop that is impossible to escape if our only thought is to make driving more convenient.

Most of the professional planners[31] and citizens would have been more skeptical of the Cross Bronx Expressway if they had known that building it would destroy neighborhoods. There's no going back in the Bronx, but there is a bright side elsewhere[32] as lessons from the Cross Bronx Expressway are used for ammunition to shoot down other projects that would ruin inner cities—without improving traffic.

Flows are one reason why systems with the exact same elements can have totally different behaviors. Examples are a highway that cuts across New York City and the soccer tactics that made "Remember Istanbul" a phrase guaranteed to make Liverpool supporters smile.

29. Proposed amounts for a carbon tax range from tens to hundreds of dollars per ton and are projected to increase over time.

30. Economist Anthony Downs coined the phrase "fundamental law of road congestion." Since then, countless studies have shown that adding roads doesn't permanently relieve traffic. This article is a comprehensive overview: Gilles, D., and Turner, M. (2011), "The fundamental law of road congestion: evidence from U.S. cities," *American Economic Review*, 101(6), 2616–2652.

31. Robert Moses, the powerful New York City planner behind the expressway, would rationalize away negative impacts to local neighborhoods with arguments about greater good for commuters (which turns out to not always be the case).

32. In the immediate vicinity of the Cross Bronx Expressway, urban visionary and activist Jane Jacobs led a coalition of citizen groups who worked throughout the 1950s and '60s to successfully stop a proposed expressway through lower Manhattan.

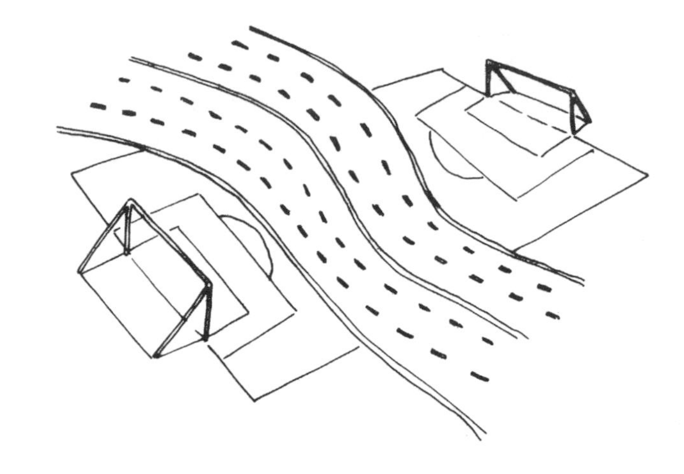

3

SMOKING MANAGERS AND ELECTRIC-BILL PEER PRESSURE

Information Flows

César Luis Menotti smoked cigarettes during games when he managed Argentina to victory in the 1978 World Cup. Enzo Bearzot preferred a pipe when he managed Italy to World Cup glory four years later. The Dutch mastermind Johan Cruyff reacted to a cancer scare by switching from cigarettes to lollipops to satisfy his sideline oral fixation.

FIFA, soccer's international governing body, has since figured out that allowing managers to smoke on the sidelines doesn't set the best example for impressionable young viewers.[33] Flouting FIFA's ban on sideline smoking, Mexico's manager, Ricardo La Volpe, lit up repeatedly as his team played Iran during the 2006 World Cup. FIFA gave La Volpe a warning.

Even when they don't have cigarettes, pipes, or lollipops sticking out of their mouths, soccer managers appear calm on the sidelines, especially when compared to coaches in other sports. American football coaches seem to be landing fighter jets with their hand signals and headsets, which they use to communicate with players on the field and with extra coaches sitting high above the field to get a better view. Basketball coaches call plays and time-outs to impose their will on the game. I enjoy the ones who move around so much that they end up sweating through their expensive suits. These seemingly bizarre actions are just

33. FIFA doesn't allow managers to smoke on the sideline because it sets a bad example. FIFA also forced every World Cup stadium in Brazil to sell alcoholic beverages from Budweiser, a company that had paid plenty to be the official beer of the 2014 event.

coaches trying to facilitate flows of information in the American football and basketball systems.

Information flows shape the soccer system too, and many of these flows are even aided by the managers—just in less obvious ways than time-outs and headsets.[34]

Managers are paying attention to information flows well before the game even starts. They instill team goals and mindsets in practices and other team functions. They use incentives and rewards, the most direct being playing time. They select and teach tactics to draw out the strengths of individual players within the team's overall strategy. Leading up to games, managers share more specific tactics, including players' roles on some of the plays in soccer that often lead to goals: corner kicks and free kicks. When defending the latter, my broad shoulders and suspect defensive ability meant that I was often selected to serve in human walls, which require no skill—just a blend of size, courage, and naïveté.

Leading up to specific games, soccer managers collect and synthesize information on the opponent. This includes everything from team tactics to strengths and weaknesses of individual players and their tendencies on penalty kicks.

Immediately before the game, the manager reviews tactics and offers last-minute inspirational words. The same motivational tricks get stale, so managers constantly develop new material, which leads to some strange speeches. One manager tried to excite us by telling us that scoring a goal is almost as good as sex.[35] Another manager explained how he used to "get laid." I'll never forget that part of his speech—but if he made another point that was supposed to inspire us, I have no idea what that point was.

Once the game begins, soccer managers have fewer opportunities to stamp their imprint on a game than coaches in other sports. Time-outs are not allowed; neither are headset connections between managers and players. Managers can try shouting or using hand motions to make changes, but doing so can distract players, since there are so few breaks in the action. Halftime is really the only in-game opportunity for soccer managers to make significant adjustments, as Liverpool's Rafa

34. Information flows in the soccer system are not limited to those between manager and player. When players fall down after not being touched or make injuries appear worse than they really are, they are manipulating information flows back to the referee.

35. I contend that a big goal can be even better than sex.

4

Benítez did to upend that 2005 Champions League final. "Remember Istanbul!"

Information flows stimulate responses in elements, physical flows, and system behavior. Flows of information can be hard to see because there may not be any corresponding physical movement.

Soccer teams' success or failure is based on how their managers deal with flows of information. Similarly, sustainability performance also depends on how information flows are managed. I see one of my favorite examples every month on our home electricity bill.

For the past few years, our bill has displayed a simple graph showing how our electricity use compares with the use of similar households in our area. There is no moral lesson on the bill about how it's wrong to use more than our share; the power company simply provides information in a new format, which allows us to draw our own conclusions.

More and more power companies are adding this comparison information to bills, because doing so leads to large and sustained[36]

36. There can be a small rebound effect from programs that show us how we are doing compared with our neighbors. It happens because those who use the least amount of electricity think, "We're doing quite well comparatively, maybe we should use more." But this effect is tiny compared to the overall reduction in use.

reductions in energy use. Using less electricity saves customers money and cuts down on the climate-changing emissions that come from the nonrenewable sources of energy that feed most power plants. But don't the power companies make more money the more electricity we buy?

Yes, up to a point. But the power companies also have a financial incentive to reduce use in the increasingly common scenarios where spikes in energy demand must be met with more costly backup generation sources or, in extreme cases, can entirely overwhelm the electric grid and lead to blackouts.

So, by identifying information flows and aligning them with sustainability goals, power companies can effortlessly save energy. And by attending to information flows before the soccer game begins, Johan Cruyff could save his voice and enjoy his lollipops.

Information flows, just like physical flows, connect system elements and influence system behavior. Information flows explain why the layout of electricity bills sways how much we use—and why, as a manager, Johan Cruyff sucked lollipops on the sideline.

PORTO/CHELSEA AND A RUSSIAN "OIL"GARCH

Stocks

If you had to choose one or the other, which of the following scenarios would you prefer for your favorite professional soccer club:[37]

A shockingly rich owner who can buy high-priced stars?

Or an established program to find and develop young players who are less expensive?

Buying high-priced stars means you don't have to wait for young players to develop and will probably see results on the field sooner. But, as the special[38] Portuguese manager José Mourinho said of this approach,

37. If you don't have a favorite professional soccer club, you should get one. You are welcome to hop on the FC Barcelona bandwagon with me.

38. Mourinho gave himself the nickname "The Special One." He's not modest, but he is brilliant. I was scared when Real Madrid (Barcelona's main rival) hired Mourinho, and thrilled when they sent him back to Chelsea.

"It's an uncertain project. It's interesting for a coach to have the money to hire quality players but you never know if a project like this will bring success."

In this free-spending-owner scenario, the owner's money is a stock, as are the players already bought with that money. Physical flows (such as players bought from and sold to other clubs) and information flows (such as evaluation of players) interact with these stocks and can change them.

While buying the stars brings instant gratification, focusing on development may work better over a longer period. The young players will improve enough to contribute, and more players like them will be coming up right behind. This larger pool of players helps the financial bottom line too, because the club can sell those players it considers replaceable.[39]

Using the youth-development approach to build their stock of players, the Portuguese club FC Porto won the 2004 Champions League. In the final game, nine of the eleven players who started for Porto were homegrown Portuguese players. The team was led by their special Portuguese manager, José Mourinho.

Less than forty-eight hours after Porto's 2004 Champions League triumph, Mourinho was hired away to manage the English club Chelsea FC. It took eight years, and eight different managers,[40] but Chelsea eventually won the Champions League using the rich-owner-and-expensive-player(-and-manager) approach. Chelsea's championship-game starting lineup included established stars from seven countries.

Stocks are built-up elements and flows. Flows are usually verbs (sell, buy, plunder), whereas stocks are usually nouns (player, money, oil). Stocks show what has happened, to date, in the system because they create memory of flows.

39. Many professional clubs treat player stocks the same way we invest in financial stocks. If another club is willing to pay more than what the player/stock is worth, it makes sense to sell. If a player/stock is available for less than they are worth, it makes sense to buy. Of course, lots of factors determine a player's value. When Major League Soccer in the United States signed the English superstar David Beckham at the end of his career, he was paid more than anyone else because of his marketing appeal, not simply his playing ability.

40. Mourinho lasted at Chelsea for three years and then was followed by Avram Grant from Israel, Luiz Scolari from Brazil, Ray Wilkins from England, Guus Hiddink from the Netherlands, Carlo Ancelotti from Italy, André Villas-Boas from Portugal, and Roberto Di Matteo from Italy. After winning the Champions League, Di Matteo was dismissed a couple of months into the following season.

To see why stocks matter for sustainability, we don't need to leave the world of soccer. In fact, we can just learn more about Chelsea.

In 2003, the year before they hired Mourinho, Chelsea were purchased by Roman Abramovich, a Russian proud of his reputation as the world's biggest spender on luxury yachts. At Chelsea, Abramovich immediately began lavishing money on soccer players too, reportedly losing about a billion dollars[41] in his first eight years in charge of the club.

Abramovich has money to spend on extravagant pastimes because he acquired controlling interest in an oil company when assets that had been owned by the Russian people were sold to private industry. Abramovich quickly turned his initial $100 million investment into billions. Abramovich later admitted in court that he had paid bribes to Russian government officials to acquire this original controlling interest at a price far below what it was worth. Draw your own conclusions, but in my book, that is called stealing.

Providing the energy to meet our needs and wants is a constant and urgent challenge. Nonrenewable energy sources such as oil, coal, and natural gas are *stock* limited. No more of these resources can be made, at least not at anywhere near the rate at which we are using them. When we depend on stock-limited resources alone, we are spending our savings without having an income.

The stock-limited nature of fossil fuels is why they are so valuable, which is why Abramovich is rich,[42] which is why Chelsea won the Champions League in 2012.[43]

Renewable energy resources, on the other hand, are *flow* limited. Wind and solar power depend on the amount and frequency of wind that blows and sun that shines—and on how much of the resulting energy we are able to collect and use.

Sustainable systems balance stocks and flows. We want our favorite team to have money for expensive superstars *and* a strong development

41. This seems like a big waste of a billion dollars, but it's only a fraction of what we have spent to dig around in the pristine Everglades and eventually realize that their original condition was best. At least Abramovich received entertainment value for his money.

42. Abramovich is not alone as a soccer owner who owes his wealth to fossil fuels. Lamar Hunt, who has done as much as anyone to establish professional soccer in the United States, also gained his fortune through the oil industry.

43. Without these stock-limited nonrenewable resources, it's hard to imagine that Russia (2018) and Qatar (2022) would be hosting the next two World Cups.

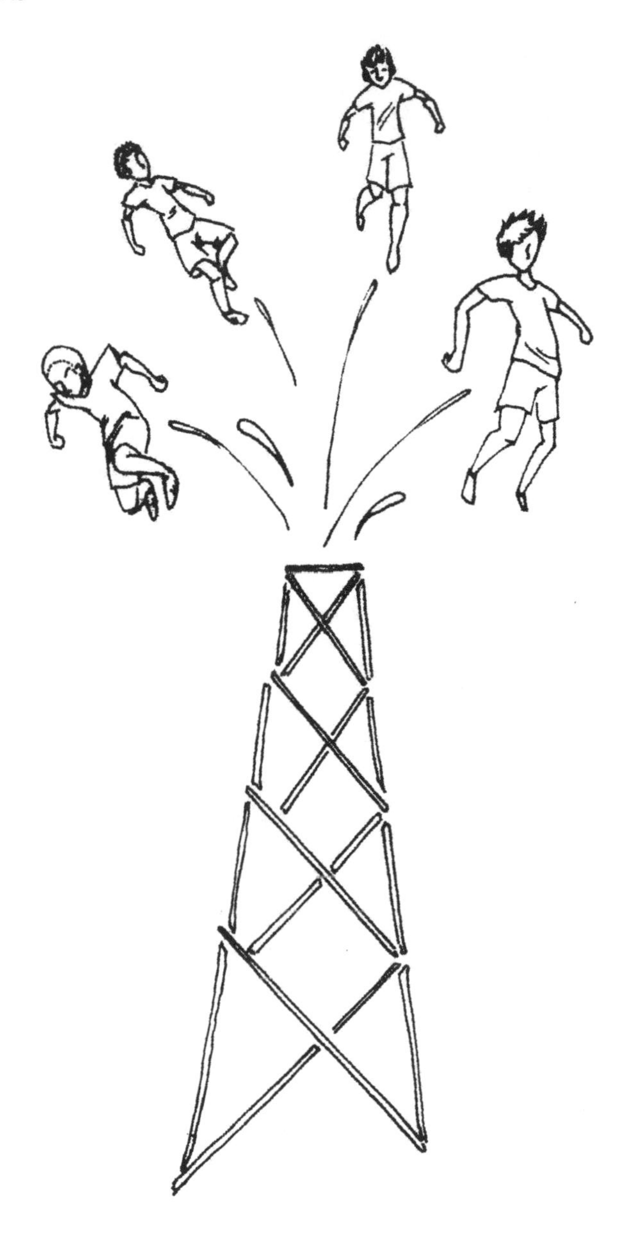

5

program producing a consistent flow of lower-priced players. Porto uses some of the money they make from selling players to buy higher-priced stars who can immediately help the team. After years of learning on the job at Chelsea, Abramovich diverts more of his oil money to player scouting, which supports the club's consistent success.

Like well-run soccer clubs, sustainable energy policy requires a balance of stocks and flows. While it is not feasible (yet) to rely solely on renewable energy flows, it's suicide to devour our nonrenewable stocks with no plan for the future.

Stocks are built-up flows within a system. FC Porto has a stock of talented young players, while Chelsea FC has a stock of oil money to spend on established players. To see stocks related to sustainability, we need look no further than Chelsea's billionaire owner, who fleeced his fortune from Russian resources.

OBSESSIVE SHOOTING PRACTICE AND POPULATION BOMBS

Feedback Loops

I spent my adolescence wearing out soccer balls by kicking them hard against things. I once popped a ball on a shot that hit the outside corner of a square crossbar; a few inches lower and to the left and it would have been just another goal. Another proud moment was when I finally cracked a board that my dad nailed up to protect the window in our barn loft from my winter indoor shooting practice. My brother obliterated the board a few years later, but that's because the structural integrity was never the same after I cracked it.

Anyway, a cycle developed around my obsessive shooting practice:

1. The more I practiced shooting, the better I shot in games.
2. As I shot better in games, I got more opportunities to shoot.
3. Getting more chances to shoot (and being better at shooting) meant that I scored more goals and my teams won more games.
4. Scoring more goals and winning more games motivated me to continue practice shooting, which brought me back to step 1.

My ball-popping, board-cracking shooting ability was born from a reinforcing feedback loop. My shooting ability was a stock; the passes I got and the goals I scored were actions dependent on the level of the stock; my practice shooting was a flow; and my improved shooting ability was a change to the stock.

Feedback loops connect a stock, actions dependent on the level of the stock, and a flow to change the stock. Balancing feedback loops slow down processes; reinforcing feedback loops speed them up.

Reinforcing feedback loops do more than build soccer skills; these loops also offer challenges and opportunities for sustainability.

Consider reinforcing feedback loops tied to human population growth.[44] More births lead to more people, which lead to more births. When the birth rate is higher than the death rate, this reinforcing feedback loop produces exponential growth. As long as basic needs are met, that's no problem; more people just mean more creativity, innovation, love, and soccer.

The challenge is that population growth increases demand for other resources. If these other resources are stock limited, as is the case with land and fossil fuels, exponentially growing demand will quickly deplete the stocks. Depleted resources lead to poverty, which leads to higher levels of illiteracy, which limits access to birth control and family planning, which further reinforces the population feedback loop.

Eventually, violent conflicts over scarce resources will control this exponential feedback loop, increasing death rates so that the population returns to a level that can be supported by available resources. Survival-of-the-fittest solves the population issue but misses the mark on sustainability, because a higher death rate is pretty much the opposite of the "meeting the needs of the present" part of our sustainability definition.

Interventions, whether intentional or not, can introduce balancing feedback loops, which slow down a process and promote stability. As we have (hopefully) learned from our Everglades experiments, ecosystems often work behind the scenes to provide these controlling services, regulating variation so that internal conditions remain stable and relatively constant.

Or, in soccer terms: my shooting didn't improve much when I was a professional, because I fell into a balancing feedback loop. Other players, like my Brazilian friend Ze, were better than me at getting into positions to score goals, so I received fewer passes in position to shoot, so I didn't practice shooting as much. When it was part of a balancing feedback loop, my shooting ability plateaued.

44. Climate change has its own reinforcing feedback loops, which is one of the reasons it is difficult to predict, and why some of the effects have been worse than anticipated. For example, white Arctic ice reflects nearly all of the sunlight that hits it, while the dark ocean reflects almost none. So warming temperatures that melt ice also lead to more absorption of solar heat in the Arctic, which means even less ice, which means even more heat absorption, which melts even more ice, and on and on.

Rather than seeking increased death rates, more civilized interventions to address the population challenge try to use stocks of clean water, food, and energy more efficiently. But using stocks more efficiently is only sustainable if the efficiency gains are exponential, just like the population growth.

Another way to bring reinforcing loops under control is through flows. China limits its population with rules restricting flows in the number of children per family. Even without restrictive rules, we could curb exponential population growth by making birth control more widely available—or just by making it more socially acceptable.

Feedback loops are when a stock, actions dependent on the stock, and a flow to change the stock are linked. Reinforcing feedback loops accelerate change; balancing loops stabilize. Examples are population trends and my soccer shooting ability.

FALLING BALLS IN BALTIMORE AND HAPPY BHUTANESE

Purposes

I was scared at first.

Then I just felt ridiculous after jumping out of the path of a car-sized metal soccer ball that was dropped from the dusty rafters of the Baltimore Arena.[45]

It was one of the first (and only) games I got to play in in my Eusébio-like[46] professional indoor career. As the visiting team, we had been announced first and were waiting on the turf as the lights were dimmed, smoke was pumped in, and the players for the hometown Baltimore Blast were introduced. The massive soccer ball falling from the ceiling was all part of the pregame entertainment. It was attached to a cable and didn't end up crushing me, stopping about ten feet above my head.

The big falling ball was more evidence for me that the indoor and outdoor professional soccer system, while it looked and felt like the unpaid soccer systems I had spent much of my life playing in, is a system with a totally different purpose. Other events that led me to this realization include, in no particular order:

the furry mascot consistently earning the loudest cheers, especially the time a shot missed the goal and went into his doghouse;

parachuters landing on the field with the official game ball (at least it wasn't a Jabulani);

stadium speakers blaring Lou Bega's overplayed song "Mambo No. 5" while I was trying to concentrate on the game;

players selecting shoe colors and crafting hairstyles to make themselves stick out for fans watching from the stands or on TV; and

me seriously wondering whether these distinctive hairstyles led to more playing time.

45. The arenas in that indoor league were unforgettable. Our home field in Harrisburg had dirt underneath the turf, which was rolled up to make way for the annual farm show. Sometimes the cows had just as many fans as we did. Detroit's Palace of Auburn Hills (an old basketball arena) has the highest showers I have ever seen. I got to play in another legendary basketball arena, the since demolished Philadelphia Spectrum, in front of sixteen thousand kids who got the day off from learning to attend one of our games. It was a bizarre experience because their cheering was completely out of sync with action on the field.

46. Just in case you've forgotten, I had just as many goals (one) and appearances (five) as the Portuguese legend Eusébio did in North America's professional indoor league. Eusébio had at least six hundred more goals than me in other leagues.

I don't know why I had to have so much evidence. Of course the main goal of the *professional* soccer system is to make money. The rules of the Game stayed the same, but the new purpose made the system behave differently from what I was used to.

If I wanted to get paid for playing soccer, then people had to be willing to pay to watch me play soccer. Or at least they had to be willing to pay to watch me sit on the bench—as my friend shot the delivered-by-parachute ball into our furry mascot's doghouse.

Back in college, where my soccer system was supported by tuition dollars, the purpose was mostly to win the game, regardless of the entertainment value for spectators, many of whom were the ones paying those tuition dollars.

In youth soccer, player development was the guiding purpose. As the level of competition increased and seasons progressed, the purpose of winning became more influential—especially for us players. Looking back, I see that the adults in the soccer system were driven to help us become better people—and we gave them plenty of ways to do so. Often, the adults' nobler purpose aligned with the purpose of simply winning games. But when the two purposes were at odds, the adults would instill lifelong lessons even if it meant temporary losses on the field.

Purposes are what a system is trying to accomplish. They have the potential to shape system behavior more than elements, stocks, flows, or even feedback loops.

Because purposes are so powerful, changing them is often an efficient way to move system behavior toward our sustainability goals. One of my favorite examples is from Jigme Singye Wangchuck, who became king of Bhutan[47] as a seventeen-year-old in 1972 and served for more than three decades before abdicating the throne to his eldest son.

Wangchuck instilled a simple but powerful shift in the purpose of his rapidly developing nation. Upon becoming king, the teenaged Wangchuck immediately required that development activities be judged on the basis of how much happier the Bhutanese people were as a result of the activity. Over time, "gross national happiness" became an official indicator of progress for Bhutan, and therefore a system purpose.

A shift in purpose that started with the young king in Bhutan is spreading all over the world. More and more nations are paying attention to new measures of development such as the Happy Planet Index, which measures the "extent to which countries deliver long, happy,

47. Bhutan is a small country in South Asia that shares borders with China and India.

7

sustainable lives for the people that live in them." In Bhutan, four decades after Wangchuck changed how development activities were judged, people are happier than those in neighboring countries and in countries, including France and Germany, with much grosser domestic products.[48]

Making money changed my soccer experience, and measuring happiness changes accounting. We are comfortable measuring and interpreting gross domestic product, but that measure rewards spending on wars and jails and ignores some of our most cherished assets, like trees filtering the air we breathe and the vitality of our communities. And gross domestic product does not account for the fact that, beyond a certain point, affluence does not correlate with other measures of well-being.

48. The following article is a review of the academic literature on well-being and affluence: Diener, E., and Biswas-Diener, R. (2002), "Will money increase subjective well-being?," *Social Indicators Research*, 57(2), 119–169.

As Brazil's soccer-playing philosopher put it: "Beauty comes first. Victory is secondary. What matters is joy." You may disagree with Sócrates, but surely you agree that we need to measure what matters. And, as a young king in Bhutan has shown us, the simple act of *trying* to measure what matters can shift purposes for more sustainable systems.

A system's purpose is usually both invisible and the basis for system behavior. Purposes explain why people in Bhutan are happy and why I thought my life would end in front of a few thousand Baltimoreans.

Review: Parts

Elements are the building blocks for the system and visibly shape system behavior.

We tend to overlook, or even intentionally manipulate, certain system elements—until our negligence is revealed through problems in a more noticeable part of the system.

Physical flows are movement over time. These flows modify behavior of the elements, and therefore of the system.

Information flows stimulate responses in elements, physical flows, and system behavior. Flows of information can be hard to see because there may not be any corresponding physical movement.

Stocks are built-up elements and flows. Flows are usually verbs (sell, buy, plunder), whereas stocks are usually nouns (player, money, oil). Stocks show what has happened, to date, in the system because they create memory of flows.

Feedback loops connect a stock, actions dependent on the level of the stock, and a flow to change the stock. Balancing feedback loops slow down processes; reinforcing feedback loops speed them up.

Purposes are what a system is trying to accomplish. They usually shape system behavior more than elements, stocks, flows, or even feedback loops.

3

Boundaries

Our approach to sustainability depends on how we define our system boundaries. What must we consider? What can we leave out? And what can we reasonably infer from the perspective we define?

FROM DROUGHTS TO FLOODS IN MAPUTO AND CAPTAIN SACRED MONSTER

Overview: System Boundaries

Systems-thinking for sustainability requires us to expand our views. That sounds simple enough. But it's a delicate task to consider enough breadth and detail that we don't miss anything, but not so much that we waste precious time analyzing things that don't ultimately matter.

What do we include? And what can we leave out?

To answer these questions, we need to remember the definition of *system:* a set of interrelated elements organized to serve a purpose. Therefore, anything that contributes to that purpose should be included in our thinking. Anything that does not can be either ignored altogether or treated as an *input* or *output,* which are ways that systems interact with their surroundings.

A focus on the system's purpose provides a rule-of-thumb starting point, but questions remain. We need to decide where to set *space boundaries,* which define the area we will consider in our systems. Different space boundaries reveal different system behaviors. The borders that define Mozambique were, as is the case for most nations, established to serve military and commercial purposes, rather than to align with historical settlement patterns or natural separations such as watersheds.

A space boundary for evaluating Coluna, the Sacred Monster, could be his play on the field. We can watch Youtube videos to see Coluna lead Benfica to two consecutive European Championships, in 1961 and 1962. We can watch highlights of when he helped Portugal knock Brazil out of the 1966 World Cup.

But, to know the full greatness of the Sacred Monster, we also need to account for the fact that he was the captain of those teams. The best captains develop and maintain team harmony and cohesion.[1] Their influence extends far beyond the playing field: from locker rooms to buses, planes, restaurants, and hotels. António Simões, a longtime teammate of Coluna, described the aura of the Sacred Monster: "His eyes alone talked to you. Both on the pitch and off it, Coluna was an example for the others. He was like a father at the head of the table. He didn't even have to speak for everyone to understand how they should behave." Clearly, if we hope to fully appreciate the Sacred Monster, we need to define a system-space boundary beyond the soccer field.

We also need to determine the *level of detail* that we will consider. Ideally, we can accurately represent the system without wasting effort and without distracting from why we're interested in the system in the first place. For an example of why it's important to get the level of system detail right, let's consider the challenge to provide sustainable access to clean water in Maputo. We'll get nowhere if we get buried in details about pumps and pipes to distribute water. That's because in Maputo, as in developing cities around the world, the rapidly growing population is concentrated in slums—without access to the pumped and piped water.

What's more, the challenge of providing sustainable access to clean water is not simply about the amount of water. In fact, too much water can also be a problem. When floods overwhelm water treatment facilities, clean and dirty water mix, and waterborne diseases like cholera spread. Floods in cities are like submerging your bathroom in a swimming pool: the sink and toilet are no longer separate water systems.

We need to answer questions about space and level of detail when defining our systems, and we also have to think about *time frames,* or the period for which we will analyze system behavior. Because sustainability is a behavior over time, we risk missing the point if we take a view that is too short.

1. Not all captains contribute positively to team chemistry. Prior to the 2010 World Cup, John Terry's affair with his England teammate's girlfriend dominated tabloids and divided the team. "Captain-for-life" John Harkes was dismissed from the U.S. team right before the 1998 World Cup.

For example, there are plenty of ways to prevent human waste from contaminating drinking water, and some of these approaches, such as raised pit latrines,[2] work better than others in flood-prone areas. By considering a time frame beginning with Mozambique's independence, we realize that Maputo has flooded in 1977, 1978, 1985, 1988, 1999, 2000, 2007, 2008, 2013, 2014, and 2015. When we project that trend into the future, it's clear that floods must be accounted for in plans to provide sustainable access to clean water in Maputo.

A longer time frame can also tell us more about Coluna's role on Portugal's 1966 World Cup team. Six years before that World Cup, the teenager Eusébio moved from Mozambique to Portugal and was plunged into an entirely new culture. The future legend made the transition thanks in part to a letter from his mom to Coluna. In the letter, Eusébio's mom asked the Sacred Monster to look after her son because Coluna was accustomed to life in Portugal, having moved from Mozambique several years earlier. Befitting his role as captain and "father at the head of the table," Coluna helped Eusébio with everything from setting up a bank account to learning the intricacies of Benfica's playing style.

Level of detail and boundaries of time and space shape our thinking and the actions that follow. Still, no matter how detailed or vast, our systems-thinking will never predict with certainty. And that's all right with me. There's no thrill to a soccer game, or to life for that matter, when we already know what is going to happen.[3]

"EL LOCO" HIGUITA AND THE NINE-DOTS PUZZLE

Space Boundaries

The aptly nicknamed René "El Loco" Higuita is famous for his Jheri curl, for his connections to Colombian cocaine kingpin Pablo Escobar,[4]

2. In flood-prone areas, raised pit latrines are sometimes used to contain human waste. Because pit latrines are dug into the ground, they can be easily overwhelmed by flooding. Raised latrines are less susceptible because they are built into mounds or structures above ground level.

3. In rare cases, knowing who will win doesn't remove all the thrill from watching soccer. During the 2004 European Championships, my friend's dad, who is from Portugal, took off from work to watch Portugal defeat Holland in the semifinals. That night, he and a bunch of friends, all of whom already knew Portugal had won, nervously watched the television replay and then recelebrated when the result was exactly the same.

4. Escobar funded Higuita's Medellín-based professional club and mingled closely with the players. When Higuita used his relationship with Escobar to arrange the release of a kidnapped child, the goalkeeper watched the 1994 World Cup from jail.

and for his goalkeeping exploits. El Loco also conceived the scorpion kick; instead of simply catching the ball as it came within his reach, he would dive forward and swing his legs up behind himself, using the undersides of his feet to kick the ball away. The scorpion kick[5] was a needless risk, but it sure was more fun than if El Loco had simply caught the ball.

Higuita got away with the odd scorpion kick because he also brought substance to his goalkeeping role. El Loco was the last line of defense for Colombia's critically acclaimed national teams of the early 1990s. He even scored forty-one goals from his goalkeeper position.[6] Or put another way, Higuita scored more than ten times as many professional goals as me, and I had no goalkeeping responsibilities.

Higuita scored dozens of goals and stopped thousands. He introduced the world to the scorpion kick. But he is most remembered for an entirely different play.

In the 1990 World Cup, Colombia advanced through their group to the sixteen-team knockout stage, where they seemed to draw a favorable matchup against the underdog Cameroon. Late in the second half of that game, with Colombia unexpectedly behind and pressing to score a tying goal, El Loco was dribbling the ball near midfield, well beyond the area traditionally guarded by goalkeepers. The risky maneuver was nothing new for Higuita, simply a calculated gamble to try to help his team on offense. But on this day, on the biggest stage, El Loco took a tentative dribble and was promptly relieved of the ball by Cameroon's substitute striker, Roger Milla, who easily scored into the unguarded goal to give Cameroon the 2–0 lead and send Colombia home from the tournament.

You couldn't script a more memorable goal. By beating Colombia, Cameroon became the first African nation to advance to the World Cup quarterfinals.[7] The goal-scorer, Milla, was thirty-eight years old, more than a decade past the typical age for World Cup strikers. What's more, Milla was only playing because of a last-minute personal request from Cameroon's president. It's not typically a good sign for your national

5. In Germany the scorpion kick is called an "Oxford," and I have no idea why.

6. Higuita's goals were from penalty kicks and free kicks, for which play is stopped. This allowed Higuita to take his time coming up the field to take the shot. But when he didn't score, El Loco had to sprint back, Jheri curl flowing in the wind, to defend against counterattacks.

7. Senegal (2002) and Ghana (2010) have since followed Cameroon's path to the World Cup quarterfinals. We'll never know how far Eusébio and Coluna could have taken their native Mozambique if it had been an independent nation when they were playing.

team when your president is selecting strikers, but it worked for Cameroon in 1990; the goal at the expense of El Loco was Milla's fourth of the tournament.

Higuita's tragedy contrasted with Milla's jubilant dancing celebration, and the goal is etched in World Cup lore.

Often overshadowed by El Loco's high-profile mistake is the fact that his playing style offered an early glimpse of what would become a positive evolution in the Game. No, the scorpion kick has not caught on. But goalkeepers do occupy positions on the field well beyond the artificial boundaries that guided tactics in Higuita's era. In today's Game, some of the best goalkeepers are "sweeper-keepers." They play further up the field and take calculated risks. They venture well beyond their goal line to intercept passes played between or over their defenders. They receive back-passes from teammates and dribble to keep possession for their team.

Higuita's unconventional approach led to changes in the Game because he defined space differently.

Whether implied or real, space boundaries determine the area we consider in our systems. How we define system space can make all the difference because different spatial boundaries reveal different system behaviors.

To reinforce this idea without the help of a Colombian goalkeeper, let's try to solve the nine-dots puzzle. Our task is to connect the dots in the figure above by drawing four straight and continuous lines passing through each of the nine dots. The end point of each line must be the beginning point of the next.

When I first tried to solve this puzzle, I couldn't think outside the box (yep, that's where the overused cliché comes from). To solve the nine-

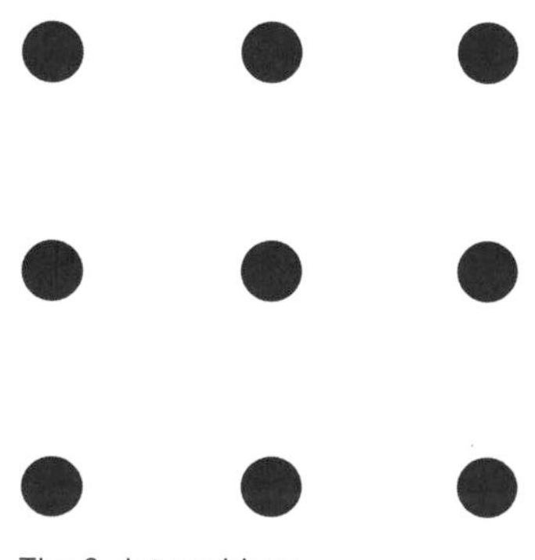

The 9-dot problem

dots puzzle, we need to draw outside the imaginary confines of the square area defined by the outer dots.

We have trouble with the nine-dots problem when we infer spatial limits to where we can draw our four lines. It's just like how goalkeepers in Higuita's era were constrained by imagined limits to where they should roam on the field.

Similarly, how we define space will guide our pursuit of sustainability. Do we consider a city? A state? A region of states? Maybe natural boundaries, such as a watershed or ecosystem or biome, would better define our system.

Defining spatial boundaries appropriately can show us things we might otherwise have overlooked. For example, a new office tower may be a beautiful and energy-efficient design. But the tower is not very sustainable if it is built in a location that occupants have to drive long distances to reach. Energy efficiency gains from the design of the building itself will be more than offset by the new transportation energy demands to get to it. Setting broader spatial boundaries in the preliminary design of the office tower would have led to more scrutiny of commuting distance and the associated effects on sustainability.

No matter how broadly we define space, we will not capture everything. That's okay—it's a necessary trade-off that allows us to devote more of our focus to the system's vital parts. Higuita's blunder in the 1990 World Cup was also a trade-off. It was an embarrassing but necessary step on the path to redefining what it means to be a goalkeeper.

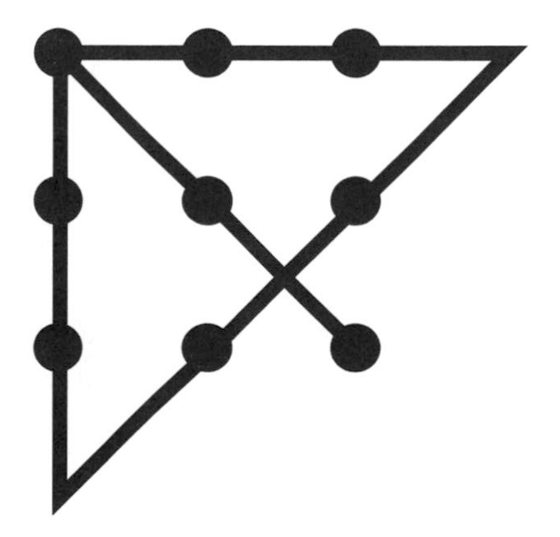

The 9-dot solution

Different spatial boundaries reveal different system behaviors. By redefining system space, we can solve the nine-dot puzzle, and a Jheri-curled Colombian can start a goalkeeping revolution.

GLORY-DAYS, SOUR GRAPES, AND SEVEN GENERATIONS

Time Frames

My senior year in high school, our team lost in the New York state championship game on a frigid November afternoon in Buffalo. The other team beat us fair and square. They were the better team that day. I would never take anything away from their accomplishments. But . . . let me explain that loss a bit more, and not (just) so I can make excuses. The loss shows that our perceptions of system behavior will change with the time frames we consider.

So, let's use what we know about the various parts of systems to review my team's loss in that championship game. As far as the most influential elements, the players, both teams were pretty closely matched. Our opponents deployed some intelligent and athletic midfielders in Sacred Monster–like roles to disrupt my team's attacking game. We still had more talent and cohesion in our tactical flows, but we may also have been a bit overconfident. We were ranked first in the state and undefeated.

Less obvious elements definitely influenced the game as well. We played on a concrete slab covered with green carpet, also known as AstroTurf. We couldn't play on grass thanks to a late-October snowstorm in upstate New York, which is a miraculously common event there and a big reason I now live where it doesn't snow. Anyway, the field was terrible, but our opponents had to deal with it too.

Nothing really sticks out until we expand the time frame beyond the ninety minutes of game time. Our game followed the championship game for smaller schools and, because all the games had been moved to the AstroTurf field, there was no space for us to warm up beforehand. No problem—the referees assured us we would have plenty of time to do our regular preparations on the field when the game before ours ended.

Apparently, to the never-touched-a-ball-in-their-life referees,[8] "plenty of time" meant about five minutes. By comparison, our usual warm-up routine took closer to an hour. As in pretty much any activity, warming up helps in soccer,[9] especially when the temperature is literally freezing.

The game started and, sure enough, we gave up an easy goal less than a minute in—the first time we had been behind all season. But again, just as both teams played on the terrible field, both teams played without the full warm-up time.

It's not until we expand the time frame a bit more that I can make my really good excuse for why we lost that game. The state semifinals were held the day before, and our opponent in the final had finished their semifinal game around two in the afternoon, twenty-two hours before kickoff of the final. That's not much time to rest and recover, but it's a vacation compared to the time our team got. Our semifinal game ended after midnight, which meant we had less than twelve hours between our previous game and the final. Not enough time to recover, but enough to need to warm up again.

After our team scarfed down late-night Buffalo wings (when in Rome . . .), entertained the groupies (not all the players did this, Mom), and smoked cigarettes (again not everyone), no one got much sleep between the semifinal and final games. The other team won the game fairly and for many reasons, but expanding the time frame lets us learn more about why.

Time frames are the period for which we study system behavior. The time frame we consider is crucial—systems behave quite differently over a minute than they do over a year or a century.

Sustainability often means extending time frames—to ensure the rights of future generations. But such long time frames present a challenge, because those whose rights we are trying to ensure may not even be born yet, which makes it impossible to get their input into present-day decision making.

Despite this logistical snag, there are still some ways to systematically consider the rights of future generations.

8. As was the case in every game I ever lost, the referees that day were terrible and completely biased against us.

9. The practice of warming up to be able to begin games at full performance was pioneered by the dominant Hungarian team of the early 1950s, which famously beat the traditional power England by a score of 6–3 and then, in the rematch, throttled them 7–1.

2

Iroquois Indians, whose confederacy used to span what is now New York State, make decisions on a seven-generation time scale—it's a requirement written into their constitution.

Japan's constitution explicitly gives rights to "the people of this and future generations."

Hungary has a government official whose job is to ensure democratic representation for the "most excluded of the excluded"—future generations.

Bolivia's constitution says that everyone has the right "to enjoy a healthy environment, ecologically balanced and adequate for wellbeing, safeguarding the rights of future generations."

Ecuador's constitution clarifies that "Nature, where the life is created and reproduced, has as a right that its existence is integrally respected as well as the right of the maintenance and regeneration of its vital cycles, structures, functions and evolutionary processes."

Even in Pittsburgh, Pennsylvania, an industrial city that I imagined as covered in coal ash until I lived there, the City Council passed a law giving Nature "inalienable and fundamental rights to exist and flourish" in the city.

Of course, even when there is no mention of future generations in governing documents, we don't intentionally make life difficult for those who follow. But unfortunately, that is often exactly what happens because of the embedded time frames in many of our systems.

Lacking a dedicated voice for future generations or an Iroquois-style seven-generation requirement, elected officials in the United States are inevitably swayed by the short-term feedback from election cycles; voting for an energy policy that is best for the next fifty years could get them voted out of office in two.

Publicly held corporations have an even shorter time frame. They have to report their financials every three months, which distracts from longer-term performance. And, for example, it takes longer than three months for gender-equality initiatives to yield the improvements that will show up in financial performance.

Soccer managers on a one-year contract have a time frame that forces them to win immediately. It shouldn't be surprising when they make decisions that are not in the long-term best interest of their team, such as trying to squeeze a few more quality games out of aging players rather than developing young ones with more potential.

Few sustainability issues are limited to such relatively short time horizons. And I'm sure it's a complicated process to go through the U.S. judicial system to officially extend human rights to future generations—but we've managed to do it for corporations.[10]

The time frames we define determine the system behaviors we see. And that's my excuse for losing the biggest game of my life up to that point. Time frames also underpin a governing idea for sustainability that the Iroquois Confederacy has always used and that present-day institutions are beginning to adopt.

A BENT-LEGGED ANGEL AND SUSTAINABLE HAMBURGERS

Level of Detail

Brazil's Pelé is the best-known soccer player of all time and perhaps the greatest.[11] Pelé's less famous countryman Manuel Francisco dos Santos

10. Since the 1800s, the U.S. Supreme Court has consistently ruled that corporations have the same rights as people under the fourteenth amendment, which forbids states from denying any person the equal protection of the laws. It seems fair that future people should have the same rights as present people.

11. Pelé, whose full name is Edson Arantes do Nascimento, gets my vote in the greatest-of-all-time debate. His three World Cup titles and over a thousand professional goals are unmatched.

also has some devout supporters in the impossible but captivating greatest-ever debate.

Dos Santos, also known as Garrincha, also known as "the joy of the people," also known as "the little bird," also known as "the bent-legged angel," was an instrumental attacking player when Brazil won their first World Cup in 1958. Four years later, with Pelé injured and unable to play, Garrincha scored more goals than anyone at the 1962 World Cup and Brazil won again.

I could go on about Garrincha's soccer accomplishments,[12] but that would never capture his essence. Watch a soccer game nowadays and you are likely to hear the crowd reward skillful play with bullfighting chants of "olé, olé, olé . . . ," often when the team that is ahead strings together a series of passes while their opponents helplessly chase after the ball.

Well, legend has it that way back in 1958, Garrincha inspired the very first olé chants at a soccer game. What's more, the bent-legged angel did it all by himself, without passing to any teammates. Apparently, Garrincha dribbled past an unfortunate Argentinean defender,[13] then allowed the poor guy to catch up, just so he could dribble past him several more times, all to the delight of the crowd.

Others have scored more goals and won more games. Garrincha earned his place in the greatest-of-all-time debate with deft skills and his carefree approach to playing the Game.[14]

Garrincha's contributions to soccer lore are astonishing when we learn that he had birth defects that left him with a deformed spine, a right leg that bent outward, and a left leg that was six centimeters shorter and curved inward. Fortunately for soccer fans and for Garrincha, his skillful play more than offset any limitations from his seemingly less-than-perfect physical attributes.

But Brazil's attempt to measure mental attributes when selecting their national teams nearly robbed fans all over the world of seeing Garrincha's brilliance. In 1958, Brazil's new psychological tests found that

12. Garrincha was not an angel off the field. The complicated man was an alcoholic who had two wives, multiple affairs, and at least fourteen children.

13. Brazil and Argentina's soccer rivalry is like no other. In the mid-1900s, it got so intense that the neighboring countries didn't play each other at all for about a decade, each avoiding tournaments in which the other was participating.

14. Set aside at least an hour if you start to watch Garrincha highlights on Youtube.

Garrincha had below-average intelligence and lacked the personality and aggression to contribute to a team.[15]

Brazil's psychological testing was massively flawed. Most notably, the testing didn't account for age and therefore penalized younger players, such as the twenty-four-year-old Garrincha and the seventeen-year-old Pelé, who also received low scores on the test.

Partly because of the psychological tests, both Garrincha and Pelé were left out of the starting lineup in the first two games of that World Cup. Not until after Brazil played poorly in those games were the two young stars given the opportunity to start in the third game.

The rest is history. Brazil won that 1958 World Cup in Sweden, and the next one, in Chile. In fact, Brazil never lost a game in which both Garrincha and Pelé played.

This is not to say that the attributes Brazil was testing for—intelligence, aggression, and teamwork—are unimportant in the soccer system. It's just that the detailed results from the off-the-field psychological testing were trivial compared with simpler and more accurate ways to evaluate those attributes.

For instance, a good way to anticipate how soccer players will do in a World Cup is to watch them play. Managers charged with selecting players cannot recreate World Cup games, but they get to see players compete in practice with the national team and in games with their club teams. Garrincha showed his quality, time and again, in those on-the-field scenarios.

Level of detail is how specific we get within our space and time boundaries. Our challenge is to accurately represent the system without distracting from why we're interested in it in the first place.

In the quest for sustainability, we need to consider some level of detail; otherwise we might as well just make random guesses. On the other hand, too much detail can waste precious time and effort as we overanalyze something that really doesn't matter, distracting ourselves from the broad and long views required for sustainability.

For instance, the more I learn about the environmental and social effects of our food system, the more questions I have:

15. Garrincha looks less excited than his teammates in videos of the celebrations following the 1958 World Cup final. Some say this shows his low intelligence, because he didn't realize how momentous the victory was—or perhaps thought there were still more games to play. I prefer to think that Garrincha's reaction reflects his ability to put soccer in the proper perspective, which is a level of intelligence I have yet to achieve.

Should I use the gas to drive to the farmer's market for local tomatoes, or just pick up what they have while I'm at the supermarket?

Is microwaving eggs from across the country better than frying eggs from a local farm?

Or, as my wife reminds me: "How much energy does it take to get your stupid dark chocolate shipped directly to our house?"[16]

And it's not just the environmentalists and foodies who care. Even the fast-food giants are asking themselves questions about sustainability. One such company, which I'll call McDonald's because that's its name, wants to create a market for and then buy "sustainable" beef. McDonald's recognizes that the necessary first step is to "support development of global principles and criteria"—in other words, figure out what the heck sustainable beef is.

Life-cycle assessment is a scientific approach to guide sustainability decisions like these. There's more to come about life-cycle assessment (and World Cup footprints) later, but for now we just need to know that these assessments help us evaluate the impact of a product or service throughout its life.

A life-cycle assessment for the beef would take account of how the cows are raised; how they are butchered and converted to hamburgers; how the hamburgers are transported to hamburger-eaters; how the hamburgers are stored, cooked, and disposed of; and so on.

Like player selection for the Brazilian team, life-cycle assessments can get quite detailed. It's impractical (no matter how many cars are ahead of you in the drive-through) to do a full assessment of whether a chicken sandwich or a hamburger is the more sustainable lunch choice.

For McDonald's, it's probably worth doing the detailed assessment. They sell a lot of hamburgers. But if I had to do a comprehensive life-cycle assessment every time I wanted to eat, I would have starved to death by now.

Fortunately, just as there are ways to simplify player selection in soccer, we can make educated food choices without going hungry. In most food decisions, we are choosing between different options, not from an end-

16. My answers to these questions are as follows: buy the supermarket tomatoes and save the gas; microwave the eggs and leave the stove off; and "I don't know exactly, but it has to be less energy than getting the chocolate from the store. It would have to be shipped there too, right?"

less menu. So, for example, we don't have to know the exact amount of CO_2 emissions tied to that hamburger, just how the hamburger compares to the chicken sandwich we are also considering.

Comparisons also allow us to further simplify the life-cycle assessment by just focusing on the differences in the dominant life-cycle phase. For the chicken sandwich/hamburger decision, the difference in environmental impact of transportation and preparation is negligible compared to the difference in environmental impact of raising a cow versus a chicken. Kilogram for kilogram, cows require more food (ten to fifteen kilograms of grain to produce one kilogram of beef) and more clean water (more than two thousand liters to bring just one hamburger to our plate).[17] Cows also occupy more land, produce more waste (roughly sixty-eight kilograms[18] of manure per day), and fart out way more climate-changing methane[19] than chickens.

Another way to make an educated decision without getting bogged down by details is to use what already exists. There are plenty of calculators on the Internet for comparing different food choices. There's also a decent chance a graduate student somewhere has already devoted a couple years of their life to doing a detailed life-cycle analysis for the very choice you are making. It's worth doing a quick Google search to check.

I applaud McDonald's for trying to make beef more sustainable. I expect that, as they claim, the results will reduce negative impacts within ecosystems and nutrient cycles and improve the lives of the beef-raisers and the welfare of the animals.

The only problem, and it's a big one, is that there are few things we can eat that are less sustainable than a hamburger. OK, a cheeseburger is worse. And a double cheeseburger worse still. The point is that, even compared to other meats, beef uses more land, water, and feed and contributes more to climate change. More sustainable hamburgers are slightly better, but certainly not a major part of a truly sustainable food system.

17. To see how much water we eat every day, go to www.angelamorelli.com/water.

18. The numbers on cow manure vary by type, size, and diet of the cow, among other things. We'll use sixty-eight kilograms here because that's what it says in the video about farming that my nephew watched endlessly when he was three.

19. Methane from livestock contributes significantly, about 14 percent, to worldwide greenhouse gases. Of course, blaming the cows for methane is like blaming cars for CO_2—our human activities are the underlying cause.

If McDonald's really wants a sustainable food system, they can figure out ways to reduce beef consumption, not improve its production. They can make foods that are lower on the food chain more appealing. But we probably shouldn't hold our breath waiting for a business that was built on beef to make this happen.[20]

It's not that details are unimportant; it's just that most of the insight often comes from just a few key details. A more sustainable hamburger is still a hamburger. And sure, I would love to know what went on in Garrincha's head, but whatever it was didn't stop him from being a transcendent soccer player. An angel with bent legs is still an angel.

While more details can improve decision-making ability, too many details can waste our time (like arguing about which hamburger is more sustainable). Even worse, too much detail may obscure what is really important and lead to regrettable decisions (like leaving Garrincha on the bench).

20. Am I a vegetarian? More than I used to be. As I have learned about sustainability and food systems, I have gone from eating meat at least twice a day to eating it about twice a month, and only when the meat would otherwise be wasted, like when my wife doesn't finish her pork ribs (sometimes I do encourage her to order a bigger size than I know she is likely to finish). Overall, I enjoy food more now, or at least I've convinced myself that I do.

ICELANDIC VOLCANOES AND THE BEST TEAM EVER

Inputs and Outputs

Even when there is ample time to prepare, and even when the Jabulani isn't the ball, we have plenty of scapegoats when our teams lose. There is the biased referee of course, and also the goalkeeper who let in a weak shot, the forward who missed an easy chance to score, the manager who picked that shaky goalkeeper and wasteful forward, the parents who raised that biased referee, and so on.

Talented excuse-makers (the type who still haven't taken responsibility for losing their high school state championship game two decades ago) can even rationalize losses on the basis of factors beyond the typical system boundaries. Consider, for example, the effect of the playing field. An inconsistent surface means more randomness in the game, which gives the weaker team a better chance for a lucky result.[21] A muddy field was one reason why a dominant Hungarian team that hadn't lost in five years were beaten in the 1954 World Cup final by the same West German team they had pummeled 8–3 just two weeks earlier.[22]

We even find influences beyond the playing field itself. A stadium packed with rowdy fans motivates players—and sways referees. The local climate shapes the action too. Games played at high altitudes[23] or hot temperatures[24] mean that players cannot run as much, which discourages the use of high-pressure tactics.

21. Because poor fields generally help weaker teams, dominant teams sometimes face unmowed grass when playing away. The grass on my college field was long, but this was because groundskeepers spent all their time manicuring the practice field for American football.

22. In addition to the muddy field, suspicious refereeing, superior equipment, and performance-enhancing drugs aided West Germany. The referee called back what would have been the game-tying goal by Hungary's legend Ferenc Puskás, who was playing with a fractured leg. Also, the West Germans had Adi Dassler himself sitting on their bench and were wearing Adidas's newest innovation—a shoe with screw-in studs that could be exchanged depending on the field conditions. Unlike the Hungarians, the Germans were therefore able to wear their regular shoes despite the muddy field. Finally, the German players were, according to a report released in 2013 by the German government, given an amphetamine known as *Panzerschokolade* ("tank chocolate") that had been developed for Nazi fighters.

23. Altitude is the easy excuse for all the teams that lose to Bolivia in La Paz, which sits nearly four kilometers above sea level. Never mind that most of the Bolivian players aren't used to that high elevation either because their club teams are in other countries or lower-altitude Bolivian cities.

24. Climate is among the more creative excuses for England's inability to meet irrationally high expectations in major tournaments, which are usually played in hotter climates than the English players are used to.

4

FC Barcelona, my favorite club team, is not immune to outside influences. Between 2009 and 2011, Barcelona were unquestionably the best team in the world.[25] Over those three years, they won three consecutive Spanish league titles and, in twenty games against the very best club teams from all over Europe during the Champions League elimination rounds, Barcelona lost just once.

In that loss, Barcelona couldn't overcome a disciplined Inter Milan team that played solid defense and capitalized on a few good scoring opportunities.

In that loss, Barcelona also couldn't overcome the volcano Eyjafjallajökull[26], which had erupted in Iceland, spewing clouds of ash that

25. A strong case can be made for that Barcelona team as the best ever assembled. They have my vote. Of course, arguing about the best team ever is just as impossible (and just as fun) as arguing about the best player ever.

26. Yes, Eyjafjallajökull is really the volcano's name—and no, I don't have any idea how it's pronounced.

grounded flights all over Europe. So, instead of taking their typical chartered flight, Barcelona's players had to travel fourteen hours by bus to Milan the day before their Tuesday game. The inconvenient[27] travel interrupted the players' customary recovery time following their game the previous Saturday.

Barcelona's players and manager—who said, "that's Mother Nature for you"—were too classy to use the volcano as an excuse, but I'm not. Barcelona looked tired during the game against Inter, uncharacteristically giving back an early lead and allowing three goals for the only time that season. Barcelona rolled to Champions League glory in both 2009 and 2011.[28] For me, that 2010 loss to Inter is an aberration that can only be explained by the volcano.

Regardless of where we define boundaries, systems are not isolated from their surroundings. We can treat these influences as inputs and outputs, which are ways systems interact with their surroundings.

We don't even need a separate, non-soccer, sustainability story here. That 2010 game in Milan shows how a rapidly changing climate touches every part of our lives.

No matter how broadly we define space and time boundaries for our systems, there will still be inputs and outputs from the surroundings. An erupting volcano in Iceland created an input that led to the unexpected elimination of FC Barcelona from the 2010 Champions League.

GREECE VS. EUROPE AND EHRLICH VS. SIMON

Insight, Not Clairvoyance

Sixteen teams qualified for Euro 2004, a tournament held every four years between the top national teams in Europe. Just about all of these teams were thought to have better odds to win than Greece.

The bookmakers who set the gambling odds for international tournaments aren't stupid. To determine odds, they consider things like past

27. My professional team routinely took long bus trips, and I've never once been on a chartered flight. That's the difference between being good enough that four thousand fans attend your game and being good enough that eighty thousand fans attend and hundreds of millions more watch on television.

28. In 2011 Barcelona traveled to the final earlier than they had planned, so that their chartered flight could avoid looming ash from another volcano.

performance, player health, likely fan support, and so on. They use statistical calculations with intimidating names like "negative binomial distribution" and "time independent least squares regression."

When bettors place a disproportionate amount of money on one team,[29] the bookmakers adjust the original odds and continue taking bets using the new odds. The bookmakers want to get equal money on either side of a bet, which guarantees they will make money (from their fees) regardless of the outcome.

All this is to say that plenty of thought goes into the betting odds, and nothing about those odds indicated that Greece would win Euro 2004. Before the first game of that tournament kicked off, bookmakers gave Greece a 1:150 chance to win. Put another way, the Greeks were such an underdog that I could have bet $2,000 on them before the tournament began and been paid $300,000 if they somehow managed to win.

And win they did, completing their improbable run through the tournament by beating host Portugal in the final. That's why they play the games instead of just having the bookmakers predict who should win.

But the outcomes of soccer games are not random. After Euro 2004, soccer pundits recognized that Greece won with their organized and passionate defending, opportunistic offense, and a healthy dose of luck.[30] And anyone who watched Euro 2004 could explain how the Greek defender Angelos Charisteas scored the tournament-winning goal: he ran past the player guarding him and used his head to redirect a perfectly placed pass from a corner kick downward and past the Portuguese goalkeeper.

Retrospectively, everything that happened in Greece's run to Euro 2004 glory obeyed the laws of physics; the ball didn't fly through Charisteas's head. But no one predicted these events prior to the tournament, nor could they have.

No matter how broad our system boundaries or how detailed our analysis, we can't be clairvoyant (I'm sorry). But systems-thinking can help us think through what is likely to happen if events unfold in different ways.

29. Among the causes of unbalanced betting are fixed matches, or just rumors of them, and irrationally optimistic England supporters.

30. Greece were outshot by their opponent in every game at Euro 2004. In the final against tournament host Portugal, Greece had just four shots and Portugal had sixteen. Portugal earned ten corner kicks; Greece had just the one on which they scored.

Fourteen years before Greece's finest soccer moment,[31] the environmentalist Paul Ehrlich mailed the economist Julian Simon a check for $576.07. The check settled a bet made a decade earlier—not on a soccer game, but on commodity prices.

Ehrlich, a biologist and educator, had bet that five commodity metals would become more expensive as human population and demand for resources continued to trend upward. He figured that as more metals were used, fewer metals would be left in the ground, which would make it harder to get the metals, which would lead to higher prices.

On the other side of the bet was Simon, an economist and educator. He believed that prices in these commodities would not rise because ingenious humans would find new ways to retrieve and use the metals, which would keep prices from climbing during the ten years of the bet.

You may be surprised to learn, in a book about sustainability, that Ehrlich paid up when the inflation-adjusted price for all five metals declined between 1980 and 1990, the period for which he and Simon had made their bet.

Fundamentalist Simonites misconstrue the bet's outcome to contend that, in a free society, we will always find solutions. For them, Ehrlich's check is evidence that we need not worry about any limits at all. In the bet, scarcity in the commodity metals was immediately followed by prices that were even lower than before there was scarcity. In the fundamentalist's logic, it therefore follows that all problems are actually good because brilliant human responses will end up making society better off than if the problems never arose in the first place.

One counterargument is that these are quite ambitious generalizations to draw from a bet that was limited to five metals and ten years—especially given that the worldwide economic recession in the early 1980s probably had as much to do with metal prices not rising as any human ingenuity. And even for these five metals, it turns out that Simon would have been the one writing the check in most of the ten-year periods over the past century.

The triumphs of Greece and of Simon remind us that we cannot predict outcomes with certainty; we can only estimate chances. No matter how brilliant and detailed our thinking, thinking is, by definition, a representation of reality—not reality itself.

31. I'm not worried at all about seeming silly in the future for declaring Greece's Euro 2004 victory their finest soccer moment; Greece will never top that run.

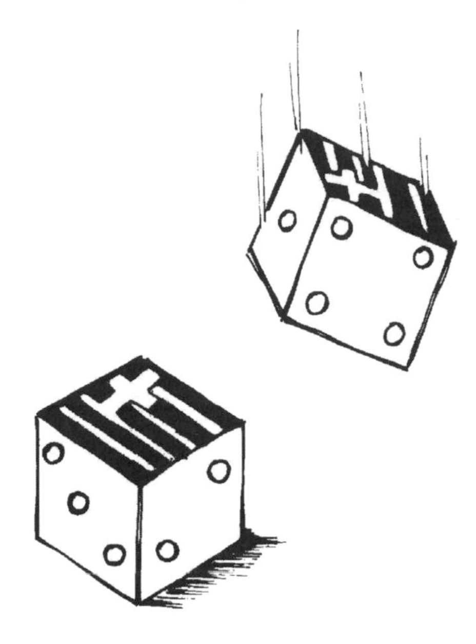

5

But enough philosophy—what does that mean as we expand our views for sustainability?

For one, even though we cannot predict the future, we can create it. On our way to doing so, we can share our more complete systems-thinking with others and refine it even more on the basis of their feedback. We can explore what-if scenarios and see possibilities. None of this will give us psychic power, but it will give us wisdom.

Plus, there are some wonderful advantages to not being psychic. The fact that we don't know what is going to happen is one reason sports, and soccer in particular,[32] are so popular. All the teams in Euro 2004 tried to give themselves the best chance to win. They developed and picked the best players. Players practiced individual skills and team tactics and studied what opponents might do. Players maintained disciplined sleep, nutrition, and sex[33] regimens. Most of these actions (there is no proof for abstaining from sex) improve the chances of winning, but none are guarantees; Greece still won Euro 2004.

32. Soccer games can turn on a single bounce of the ball—and the associated randomness that grows from it. In sports like basketball, more chances to score mean more opportunity for the randomness to get canceled out.

33. One of the least important and most publicized decisions managers need to make heading into a major tournament is how much time players will be allowed to spend with their partners.

Uncertainty is also why Simon and Ehrlich's bet got so much attention. And while neither man was positive he would win, both were confident enough to take a calculated risk.

There will always be things we cannot know, but that doesn't mean that our learning and acting to try to make things better is futile. Both Ehrlich and Simon would agree with that point, but since Simon won the bet, I'll let him have the last word here: "In this I agree with the doomsayers—that our world needs the best efforts of all humanity to improve our lot."

Our thinking, no matter where we define boundaries, cannot predict with certainty what will happen. This is why an environmentalist lost his bet with an economist and why I missed the chance to turn a month of my meager soccer salary into a nice house by investing in Greece.

Review: Boundaries

Whether implied or real, space boundaries determine the area we consider in our systems. How we define system space can make all the difference because different spatial boundaries reveal different system behaviors.

Time frames are the period for which we study system behavior. The time frame we consider is crucial—systems behave quite differently over a minute than they do over a year or a century.

Level of detail is how specific we get within our space and time boundaries. Our challenge is to accurately represent the system without distracting from why we're interested in it in the first place.

Regardless of where we define boundaries, systems are not isolated from their surroundings. We can treat these influences as inputs and outputs, which are ways systems interact with their surroundings.

No matter how broad our system boundaries or how detailed our analysis, we can't be clairvoyant. But systems-thinking can help us think through what is likely to happen if events unfold in different ways.

4

Behaviors

With parts and boundaries defined,
we can detect system-level behaviors
and discover how they affect sustainability.

A UNITED NATIONS REPORT AND THE SACRED MONSTER'S BROKEN NOSE

Overview: System Behaviors

Sustainability quests are underway in cities all over the world, including in Maputo, where people are trying everything from improving water resource management to raising awareness of climate change risks, to reducing social segmentation by upgrading slums.[1] What makes Maputo more sustainable won't necessarily work in Mexico City, Madrid, Milan, or Manila. Yet while every situation is unique, some universal behaviors shape sustainability performance in every city and in all sorts of systems.

To appreciate those shared behaviors, we need some more context for our definition of sustainability: "meeting the needs[2] of the present without compromising the ability of future generations to meet their own needs." The United Nations unveiled this definition in 1987, but only after a multilateral group[3] spent two and a half years traveling all

1. There are so many sustainability initiatives going on in Maputo that integration between them is a priority task mentioned in Mozambique's "Climate Change Assessment."

2. I think of "needs" as what must be in place for people to enjoy life. That's what motivates my quest for sustainability: letting as many people as possible, now and in the future, do things they find rewarding, which I hope includes playing soccer.

3. This group was the United Nations' World Commission on Environment and Development, which was also known as the Brundtland Commission after its chairwoman, Gro Harlem Brundtland. In their globetrotting, the commission didn't make it to

over the world gathering input from citizens and experts. Because of this collaborative approach, the group did more than just write a definition; they brought forth a shared global vision for sustainability.

This vision for sustainability has overlapping environmental, social, and economic dimensions.[4] And these dimensions have a hierarchy: we need an environment that supports human life (first in the hierarchy) in order to have functioning societies (second in the hierarchy), which, in turn, are a requirement for healthy economies (third in the hierarchy).

Events from Maputo make the hierarchy obvious. Social and economic activity in the city broke down when a cyclone on top of flooding (environmental dimension of sustainability) swamped the area in 2000. A decade earlier, Maputo's economy was paralyzed in the midst of Mozambique's civil war (social dimension of sustainability). It's a self-evident but easily-taken-for-granted truth that the economy depends on a functioning society, which depends on an environment that supports human life.[5]

Our shared sustainability vision does not require an end to growth. In fact, part of Maputo's sustainability plan is to grow, just as sustainable living systems do. The *type* of growth is what matters. On a materially finite planet, infinite material growth is impossible, so pursuing it is foolish.[6] We can, however, endlessly grow in more meaningful ways—in our knowledge and creativity, and in our ability to juggle tennis balls with our feet.

Sustainable systems grow through *self-organization,* in which behavior is driven by initial rules combined with interactions between elements. Coluna, the Sacred Monster, had his nose broken only eight minutes into the 1961 European Cup final. In response, he and his Benfica

---

Mozambique, where a civil war was underway. The commission did, however, spend a week gathering input in nearby Harare, Zimbabwe.

4. "Triple bottom line" accounting attempts to capture these overlapping dimensions. Social and environmental indicators are reported alongside traditional economic financials, providing a more complete measure of performance.

5. The late stand-up comedian George Carlin had a clever way of making the point that we depend on an environment that supports human life. Here's an excerpt: "The planet has been through a lot worse than us. Been through earthquakes, volcanoes, plate tectonics, continental drift, solar flares, sun spots, magnetic storms. . . . And we think some plastic bags and some aluminum cans are going to make a difference? The planet isn't going anywhere. WE are! . . . The planet will be here and we'll be long gone. Just another failed mutation."

6. Even a goldfish knows not to get too big for its bowl. Well, sort of. They produce hormones that, in nature, suppress the growth of surrounding fish, which helps goldfish maintain a size advantage. In a tank, this hormone builds up and suppresses growth of the goldfish itself.

teammates self-organized. For the rest of that game, instead of rushing toward the goal to connect with crossing passes and try to score with his head, Coluna stayed in a reserved position where the ball was more likely to come to his feet.

Coluna switched full-time to that more defensive position the following season when Eusébio was added to Benfica's lineup. The change wasn't dictated from on high by soccer gods,[7] but was a result of self-organization producing *emergence,* whereby entirely new behaviors arise from the simple organizing rules and resulting interactions. In natural systems, emergence leads to new species; in the soccer system, emergence led to a new position for the Sacred Monster.

Let's revisit that 1961 European Cup final, so I can tell you what happened with Coluna and so we can learn another behavior of sustainable systems. In that final, Coluna was *resilient* because he minimized damage from an unexpected event (the broken nose) and then, after surviving the first impact, he recovered. The Sacred Monster heroically scored the deciding goal on a long-range shot that he was in position for after adapting so that he could play on with his broken nose.

Self-organization, emergence, and resilience all require change. To determine which changes are sustainable, and which ones might break the system, we look for *thresholds,* which are crossed when a system moves from one state to another. Two years after Coluna's nose was broken, Benfica were again playing in the European Cup final when Coluna had his leg injured by a cynical foul. Unlike the broken nose in 1961, the leg injury crossed a threshold because Coluna was unable to continue playing.[8]

Thresholds, self-organization, emergence, and resilience all occur within a network of relationships and patterns. And that's why we need to respect *interdependence,* the reality that each part of a system gets its behavior from, and owes its existence to, its relationships to the other system parts. For an example, let's return one last time to that 1961

7. From what I've seen, the soccer gods do not seem to bother themselves with player positioning, instead preferring to intervene through less obvious strange bounces that either create or prevent scoring chances, depending on who the gods are cheering for that day.

8. Losing the Sacred Monster to injury was especially devastating because, at the time, injured players could not be substituted. So, not only did Benfica have to play without their captain, but they also played with one less player on the field, ten against eleven. Not surprisingly, they ended up losing the game.

final. For the select few who have scored a game-winning goal in a European Cup final, the moment is career-defining. But for Coluna, the goal wasn't even his most impressive feat that day. That's because after the game, in the victors' locker room, the Sacred Monster revived Benfica's club president, who had suffered a heart attack in the tense final minutes of the game.[9] The Sacred Monster saved his president, and we see interdependence between soccer and survival.

THE WORST GAME EVER AND MARTIN LUTHER KING JR. AS AN ENVIRONMENTALIST

Interdependence

The Soviet Union and Chile were only a step away from qualifying for the 1974 World Cup. The teams would play two games against each other, one game in each country, with the combined score determining which team got to compete in the tournament in West Germany.

The game in Moscow ended in a tie, which meant that it all came down to one game in Chile. Incredibly, the Soviet team didn't show up for the game. Even more incredibly, the game was "played" anyway, with fans in the national stadium in Santiago watching as reluctant Chilean players lined up and took the opening kickoff down the field to score a ceremonial goal.[10]

The Soviet team wasn't there because there had just been a coup in Chile, and after the coup, people deemed undesirable by Augusto Pinochet's new dictatorship were tortured and executed in the national stadium. The Soviet Union asked to move the World Cup qualifying game to a venue that had not recently been used for torture and executions. But FIFA refused when the delegation they sent to investigate reported that "based on what we saw and heard . . . life is back to normal."

9. Coluna waved his sweat-soaked game jersey to revive Benfica's president, Maurício Vieira de Brito. When he miraculously came to after his stress-induced heart attack, Vieira de Brito supposedly said, "My dear Benfica, the strongest club in Europe. Now dying wouldn't be all that terrible."

10. The farce of a game between Chile and no one was completely unnecessary. The rules are clear that if one team forfeits, the other is awarded a 3–0 victory. Perhaps Pinochet's dictatorship thought the staged event would be good propaganda, even though it just brought more notoriety.

So the Soviet team didn't go,[11] and Chile scored on an undefended goal to qualify for the 1974 World Cup, where they were eliminated in the first round.

Pinochet remained in power until 1990.

The Soviet regime outlasted him by a year.

Every part of a system is connected, which means that each part's behavior is shaped by relationships to the other parts of the system. So, we can study elements, stocks, flows, feedback loops, and purposes, but we cannot know how the system works without also seeing interdependencies.

As with most ideas, Dr. Martin Luther King Jr. described interdependencies more powerfully than I can. Just consider this passage in King's "Letter from Birmingham Jail": "Injustice anywhere is a threat to justice everywhere. We are caught in an inescapable network of mutuality, tied in a single garment of destiny. Whatever affects one directly, affects all indirectly."

The great American civil rights leader's words about interdependence show that he was an environmentalist,[12] even though he was assassinated years before the environmental movement officially began in his country. King knew that justice could not be isolated to a single place or issue. He understood that environmental quality was part of the fight for freedom and equality. And King acted on this wisdom by protesting unhealthy housing in Chicago and inadequate sanitation in Memphis.

If King's housing and sanitation protests happened today, we would probably consider them part of the environmental justice movement for equal rights to a clean and healthy environment. Not only do the civil rights and environmental justice movements share similar goals, but

11. The telegram to FIFA from the Soviet Football Federation explains why their team stayed at home: "Now in Chile prevails atmosphere bloody terrorism and repressions . . . National stadium supposed be venue hold football match turned by military junta into concentration camp place of tortures and executions of Chilean patriots . . . Soviet sportsmen cannot at present play in stadium stained with the blood of Chilean patriots." To me, that sounds like a pretty good reason for not playing, although there is also a theory that the Soviet authorities wouldn't have boycotted if they thought they could win.

12. Dr. King also got to the essence of sustainability when he implored people to "shift from a 'thing-oriented' society to a 'person-oriented' society."

both movements use similar tactics, such as demonstrations, picketing, and targeted political pressure.

Two linked principles guide the environmental justice movement. The first is that everyone deserves equal protection from environmental hazards and no one should be restricted to unhealthy living options, whether through legal or economic segregation. The second is that everyone should have equal access to the decision-making process that determines living conditions. For example, community representatives, and not just corporations and government bureaucracies,[13] should have a say in deciding where pollutants are released. And, as I will discuss later, studies of polycentric management show that the justification for broader participation in decision-making is not only moral but also practical.

If you're looking for a movement to join, plenty of help is needed to achieve environmental justice. Environmental inequality continues to take its heaviest toll on people in the low-income sections of cities and poor rural communities,[14] places that are home to a disproportionate number of racial and ethnic minorities.

Environmental inequality is why Hurricane Katrina disrupted life for affluent people in the protected French Quarter of New Orleans but ended life for poor people living in the neighboring and vulnerable Lower Ninth Ward. It's why subsistence farmers are among the first to have their lives uprooted by the droughts and wildfires that comfortable consumers have unleashed with our climate change emissions. And it's why rates of asthma[15] are so high among people with no other choice but to live in the most polluted, but lowest-cost, parts of cities like Houston and Los Angeles.

13. In the United States, government organizations alone have been ineffective in the fight against environmental injustice. The Environmental Protection Agency officially began accepting environmental injustice complaints in 1994; since then, hundreds of claims have been filed, though only a few have been resolved.

14. King called such places "islands of poverty." These islands are not relaxing destinations with sandy beaches, palm trees, and inflated food prices. Islands of poverty are isolated traps with degraded soil, polluted water and air, and inflated food prices.

15. Asthma is not the worst fate linked to air pollution. The World Health Organization estimates that at least seven million deaths each year result from air pollution.

King recognized that a just society requires an environment that supports human life. And we need a fair society to enjoy soccer. As Brazil's 1994 World Cup savior,[16] and now senator,[17] Romário de Souza Faria said before his country hosted the 2014 World Cup: "There are no good schools, there are no good hospitals—how can there be a World Cup?"

World Cups, or the lack thereof,[18] show the inescapable interdependencies between soccer and politics. Mussolini's fascist regime got a boost of nationalism from the 1934 World Cup, which Italy hosted and won. Argentina hosted and won the World Cup in 1978, and their military dictatorship decimated the national budget trying to project a positive image to the world.[19] Brazilian President Jânio Quadros officially made Pelé a national treasure after Brazil won the 1962 World Cup. "National treasure" looks good on Pelé's résumé, but the honor was really bestowed to ensure President Quadros's popularity among Brazilians who didn't want Pelé to leave their country and sign with a professional club in Europe.

The interdependencies between soccer and politics work the other way too. Colombia actually relinquished hosting duties for the 1986 World Cup. President Belisario Betancur explained: "The golden rule whereby the championship would serve Colombia and not a group of multi-national companies was not observed. Here we have many other things to do."

You now know that we find such interdependencies in qualifying games as well. Chile's dictatorship, and FIFA's blindness to it, helped Chile qualify in 1974. And the charade of their game against no one

16. Brazil won the 1994 World Cup, and Romário scored the most goals and was named the top player, which cemented his place among the best goal-scorers of all time. More than a decade later, Romário ended up playing in the same U.S. professional league in which I had mostly sat on the bench. He scored nineteen goals in twenty-six games and was named league Most Valuable Player—as a forty-year-old.

17. Part of Romário's political platform is to require soccer clubs to provide their players with educational options because they play "a very important role in the development of future generations." Romário is connecting sustainability and soccer too!

18. There was no World Cup in 1942, during World War II, or in 1946, in the aftermath.

19. Not everyone was fooled by Argentina's investment in the 1978 World Cup. Dutch visionary Johan Cruyff asked, "How can you play soccer a thousand meters from a torture center?" Unlike in Chile, the dictatorship in Argentina hadn't used the actual soccer stadium for a torture center, but it was close by. And planes from which still-living prisoners were thrown into the ocean took off from a nearby airport.

reminds us of the interdependency that is competition; without it, there is no fun for players or fans. Whenever I whined about an opponent, my mom reminded me of this interdependency: "Without him, you'd have no one to play against."[20]

Not only is there no game without the interdependencies of competition, but no one gets better. Chile didn't win a single game in that 1974 World Cup. And traditional soccer powers Argentina and England each fell behind during periods of isolation that prevented them from keeping up with the Game's evolution in other countries.

But interdependencies also mean that the Game can have positive impacts beyond the playing field. Brazil won the 1958 World Cup with a racially diverse team and therefore sped progress toward racial equality in Brazil.[21] And in the early 1980s, wearing the black and white colors of Corinthians, the club in which the Brazilian Sócrates starred, was a way to show support for democracy in Brazil. The common language of soccer unites new immigrants and native-born people all over the world. The Game entices an isolated American like me to learn not just about the geography of countries like Bolivia,[22] but about the people who live there.

Interdependencies between soccer and politics and between social and environmental justice are just a start. From Eusébio relying on the Sacred Monster's passes, to the Icelandic volcano that broke up Barcelona's dream season, I hope you are seeing interdependencies in all our shared stories. Because, as King said, we're all linked together, even to the soccer system,[23] through a "single garment" of biological and social relationships[24] that extends around the world and into the past and future.

20. Mom and I did usually agree that referees weren't necessarily essential.

21. The pioneer of Brazilian soccer journalism, Mário Filho, credited Brazil's 1958 team with "completing the work [the abolition of slavery] of Princess Isabel." Filho's 1947 book *The Negro in Brazilian Football* is still a sports literature classic.

22. After watching Bolivia and South Korea play to a 0–0 tie in the 1994 World Cup, my brother, sister, and I were sitting in the back of the family minivan (yes, the green Dodge) talking about juvenile things—we were eleven, fifteen, and sixteen, respectively, so it was excusable at the time. My dad awkwardly tried to raise the level of the conversation by asking, "So, where's Bolivia?" Even my mom laughed at his transparent attempt to change the subject—and the conversation just got more juvenile. Now, I'll ask my Dad, "Where's Bolivia?" whenever a conversation is deteriorating.

23. Just to be clear, King didn't specifically mention soccer; his sport of choice was billiards.

24. Mayans constantly reminded each other of these interdependencies in their traditional greeting, which translates as "I'm another you, you're another me."

Interdependency, the connections between parts, drives system behavior. It is the "single garment of destiny" celebrated by Dr. Martin Luther King Jr. and ignored by the Chilean dictator Augusto Pinochet (and FIFA).

PICKUP GAMES AND HUNGRY ANTS

Self-Organization

I still share a beautiful understanding with my former teammates. The more we played soccer together, the more I could anticipate where they were going to run, how they wanted to receive the ball, and what they were likely to do once they got it. I know my teammates would say the same thing, and they might even mention that passing to me usually led to a shot rather than a return pass. Even now, as an injury-prone participant in annual alumni games, I can still anticipate where my old teammates will be, or at least where they would have been fifteen years, twenty pounds, and three kids ago.

This beautiful understanding develops even among players in a spontaneous game of pickup soccer. Players start with a shared idea of how soccer works. They divide into equal-numbered teams and begin to spread out into positions. At first, too many players appoint themselves as creative maestros with no defensive responsibility, but most of them eventually move to other positions so that their team is balanced all over the field.

When the game starts, the shared understanding leads to predictable behavior.[25] An attacking player will dart away from the ball to indicate she wants to receive a pass into open space. The same attacker will run toward the ball to receive a pass if she has defenders on her back. A defender, if he's confident in his foot skills, will provide a passing option when his goalkeeper has the ball. In tight spaces, skilled midfielders make short passes and quickly move for an immediate return pass.

The information flows that drive the pickup game are often unspoken. A midfielder may subtly turn her hips to show where a pass should be played. Selfless players, probably not the self-appointed maestros, will run with no intention of receiving the ball, drawing defenders to themselves and opening up space for teammates.

The pickup dance doesn't last forever. Eventually, the pot-bellied guy gets tired, the goalkeeper goes home to eat dinner, and a displaced creative maestro stops patrolling the left sideline to return to his rightful position in the crowded center of the field.

But before the pickup game deteriorates, simple rules and shared understanding of the Game produce a team behavior that is increasingly complex and effective. Those players who skillfully control the ball and pass to others end up receiving more passes. Those who lose the ball to the other team receive fewer. So does the show-off who is dribbling too much. The lefty who scores on a well-placed shot gets more passes to her left foot in good positions to shoot. The fitness buff playing to cross-train for a triathlon runs back on defense to cover for the guy playing just so he can drink beer afterward. In most cases, these adjustments make the team better.

In a self-organizing system, behavior comes from a few shared initial conditions and from interactions among the parts of a system.

Flocking birds, schooling fish, and hungry ants are self-organizing systems, just like humans playing pickup soccer.

Take the foraging ants for example. Before they have found food, the ants appear to be randomly wandering, which they are. But they are also leaving an invisible trail of pheromone, a chemical that other ants

25. Pickup players' behavior is not entirely predictable, especially the behavior of those who have a different purpose than everyone else who's playing: the washed-up high school hero trying way too hard in the co-ed game where everyone else just wants to get some exercise; the happy-go-lucky player who is miscast in the serious game where players realize they aren't what they used to be, but still get joy out of trying their hardest; or the guy (me) who can't turn off his ultracompetitive streak and is slide tackling people in the just-for-fun game. My brother actually met his wife this way, practicing with the women's team during his off-season at college. She was apparently won over by his ability to fire shots (that could shatter boards) at a helpless goalkeeper.

2

can detect. When a lucky ant happens upon food, let's say a capsized ice cream cone, the ant returns to the nest while depositing a stronger pheromone trail.

Because ants have common trail-following behavior, more and more of the wanderers begin to follow the stronger pheromone trail and, as a result, stumble upon the ice cream. A reinforcing feedback loop results: more ants strengthen the trail, a stronger trail attracts more ants, more ants strengthen the trail, and so on. It's just like how the goal-scoring lefty in the pickup soccer game gets more and more passes.

When the ice cream is gone or the ants get full, fewer and fewer ants follow the trail and it eventually disappears.

Scientists study natural systems in search of such simple rules that can yield complex behavior. If you've been fortunate enough to watch ants as a child or with one, you know that, in addition to finding capsized ice cream cones, ants don't seem to get stuck waiting in lines. By studying this self-organizing knack, researchers found the "ant colony optimization algorithm," which is now applied to slash wait times for humans through improved shipping logistics, faster telecommunications, and more efficient traffic.[26]

26. For the shipping, check out the company Antoptima. And for a good overview of the telecommunications, see this unusually readable journal article: Schoonderwoerd,

Neither the pickup soccer players nor the ants rely on a commanding central authority. We rarely find a coach in pickup soccer. The queen in the ant colony, despite her title, is not a central decision-maker; she is a reproduction machine. Yet, because of the interactions that result from a few simple rules, spontaneous soccer becomes a dazzling dance and aimless ants get to eat ice cream.

Sustainable systems can self-organize on the basis of a few initial rules and interactions between parts of the system. Self-organization guides behavior in pickup soccer and among foraging ants.

MAYAN BALL GAMES AND CHIMPANZEES

Emergence

In tribute to the history of the Game, I convinced Monica to spend a day[27] of our honeymoon away from the tranquil Caribbean beach. Instead, we took a dangerous taxi[28] pilgrimage through the Yucatan jungle.

The journey was worth it (for me) as soon as we made it to our destination: a couple of well-preserved ball courts among the ruins of Coba, an ancient Mayan city. I stood on the same stone courts where, over a thousand years ago, Central Americans played with rubber balls and mounted rings as goals. The ball was symbolic of the sun and the solar system, while the court represented both the removal and renewal of life. The court was also where, as our soccer sage Eduardo Galeano puts it, "sometimes the winning captain was sacrificed, sometimes the losing captain."

Soccer-like games were played even before the age of the Mayans. Around three thousand years ago, a popular activity in Japan involved a dozen or so people trying to keep a ball stuffed with sawdust from hitting the ground—sort of like what we now call "hackysack."

R., Holland, O., and Bruten, J.L. (1997), "Ant-based load balancing in telecommunications networks," *Adaptive Behavior*, 5(2), 169–207.

27. Our trip to Coba was really only a half day, but it "counted" as a full day of doing what I wanted because Monica had spent the morning with me watching lifeless England escape against Trinidad and Tobago in the opening round of the 2006 World Cup. This was the game in which Peter Crouch scored the clinching goal and then, to my dismay, didn't use his robot celebration.

28. The road through the Yucatan jungle had three lanes, one for traffic in each direction and a third, in the middle, for any driver going in either direction who thought passing was worth the risk (our driver did). Speed bumps to signal approaching towns were perpendicular to the road surface and invisible until it was too late.

About two thousand years ago, soldiers serving China's Han dynasty entertained themselves by kicking balls into a high net. Around the same time, ancient Greeks were playing *episkyros,* a game that had field linings similar to those in soccer but that allowed handling of the ball.

Later, in eighth-century Britain, herds of people competed to reach a specified location while kicking a "ball," which, depending on the day, could be a pig's bladder or a prince's head.

In Italy, by the 1500s, a game called *calcio* had evolved and was contested between teams of thirty or more players trying to kick or carry a ball across a goal line.

While the Renaissance Italians played *calcio,* native Eskimos in Canada played *aqsaqtuk* on ice, using balls of caribou hair. On the milder beaches of present-day Virginia, Native Americans played *pasuckuakohowog,* which is translated as "they gather to play ball with the foot." That sounds like soccer, except that *pasuckuakohowog* was played with hundreds of players on a field nearly a kilometer wide and two kilometers long.

Indigenous Australians played *marn grook.* Indigenous New Zealanders played *ki-o-rahi.*

I think you get the idea. What we now know as soccer evolved—it was not simply created one day by some divine being, or even by the British.

To be fair, the British did contribute to the self-organizing principles that define soccer today. An initial set of rules, established at Eton College in 1815, morphed into the Cambridge Rules[29] by the middle of that century. Then, small variations in the way games were played led to slightly different versions of the Cambridge Rules, which led to more changes, which led to a fracture between the groups using the different rules. Eventually, in 1863, the Cambridge Rules were split into a set of rules for rugby football and another for association football, the main

29. Soccer and rugby are not the only sports that can trace their roots to the Eton and Cambridge Rules. American football, which is framed as the antithesis of soccer in either/or debates, actually shares common ancestry too. American football was born from Cambridge Rules games played between universities and colleges in the New England region. Divergences from rugby football introduced the line of scrimmage and down-and-distance rules that distinguish American football today. The Cambridge Rules say nothing about American football's over-caffeinated coaches gesturing like they're trying to land airplanes. Those came from post-split emergence in American football and remind us that emergence is not always positive.

difference being that the rugby rules allow players to carry the ball with their hands. With relatively minor changes since that 1863 split, the association football rules govern modern soccer.

Some emergent changes are incremental, like the team coalescing in the pickup soccer game. More radical changes can emerge when variations are amplified by reinforcing feedback, as happened when the Cambridge Rules produced both rugby and association football.

Emergence is when entirely new behaviors arise in a system, new behaviors that were not present in any of the parts alone. Therefore, emergent behaviors depend on system relationships.

So, yes, sustainability itself is an emergent property. And so are resilience, adaptability, and elegance, each of which I will discuss more soon.

As you may have figured out by now, evolution is an emergent behavior. It brings forth unique new species. Through natural selection, some traits arise and others go away, based on whether they contribute to an organism's survival and reproduction.

Humans and chimpanzees and bonobos all split from a common ancestor, creatively named by scientists "the chimpanzee–human last common ancestor." This happened roughly five million years ago, and since then, humans, chimpanzees, and bonobos have evolved along their own paths.

The Cambridge Rules are the last common ancestor for soccer, rugby, and American football. And, like apes of all types, soccer is ever changing. Examples we've covered, like Higuita leaving the goal line and infusions of money from "oilgarchs," are just a couple of the incremental emergent changes that make soccer today look different from soccer in the 1950s and ensure that we'll be treated to some unknowable innovations (hopefully not another Jabulani) at the next World Cup. Emergence also explains bigger changes. It's why soccer is, thankfully, different from American football and the sacrificial Mayan ball games.

Soccer may be constantly evolving, but it didn't evolve from rugby or football—and humans didn't evolve from chimpanzees. Sure, humans have some unique traits; we play soccer, and chimps and bonobos do not. But chimps and bonobos have unique traits too; they can quickly climb trees, which we do about as well as chimps and bonobos play soccer.

In other words, we are not the evolutionary pinnacle that has emerged from other ape species; other ape species are our closest living relatives. Just as you and I are mammals and primates, we are also apes.

Sustainable systems are dynamic, with new behaviors resulting from emergent properties that were not present in any single part. Emergence is why the English shouldn't get all the credit for inventing the Game, and why we all should respect apes.

UNFAIR GOALS AND LEWIS'S LIZARD

Resilience

Our team was pounding the opponent's goal with shot after shot. It seemed like only a matter of time until we scored. But then, in spite of our team's dominance, the ball took a funny bounce, the referee missed an obvious foul, and the worst player on the other team miskicked a long crossing pass, which miraculously floated over our goalkeeper's head and into the goal.

In soccer, even when everything seems to be unfolding perfectly as planned, adversity is always lurking just a play away. That's why, in my playing days, I kept above my bed a tattered note sheet with the quote "Life is 10% what happens to you and 90% how you respond to it." I'm sure the author[30] would be proud to know that his words (sometimes) kept me from swearing at bad referees.

At first, resilience is in how we minimize damage from, as the quote puts it, "what happens to" us. The other team's lucky goal put us behind on the scoreboard, but there was no foolproof way to prevent that from happening. Sure, our team could have started the game with all defenders, but then we would have had almost no chance to score ourselves. And, when we were scored on, we would have been stuck trying to come from behind with just defensive players.

We can't prevent the goal against the run of play, and we can't immediately fix it, but we can make things worse. Our captain could get a red-card ejection for yelling at the referee about the bad call. Now we are not only behind by a goal, we are down a player. Our manager could scream useless insults at the goalkeeper. The goalkeeper could do the same to the defenders. But that would just make everyone mad at each other, less confident, and still behind by a goal.

After surviving the initial impact without making things worse, resilience is in our response. Right after the ball goes in the net, does everyone hang their heads, already defeated? Or does a leader quickly get the ball out of the net and encourage teammates? Are players and coaches discussing tactics[31] to adapt to the new reality of needing to score at least one goal in the time that remains in the game?[32]

Resilience is the ability to resist damage from an unexpected disturbance and then quickly recover.

30. It turns out the quote was from the evangelical radio preacher Charles Swindoll. He wasn't the first to come up with the idea, though; nearly two centuries before, a Greek slave turned philosopher, Epictetus, pointed out that "we can't choose our external circumstances, but we can always choose how we respond to them."

31. In college soccer, "frantic city" was our plan for taking more attacking risks in the event that we were behind with fifteen minutes left in the game. Frantic city is also what we called the fifteen minutes before closing time at our bar.

32. We're more likely to recover the more time we have. When my team fell behind unexpectedly in the fourth minute in an elimination playoff game my senior year of college, we had eighty-six minutes to come back. When we tied the game and then scored a goal in sudden-death overtime, the game was over—there was no time for the other team to be resilient.

Depending on the day, my young neighbor Lewis can be anything from a biologist to an archeologist to a knight or a spy. His biologist days made him the most popular kid on his preschool playground, where he collected ladybugs from branches creeping through the perimeter fence. Lewis found so many that he shared with his classmates who didn't have time to look for ladybugs because of prior commitments pushing dump trucks or hanging from overhead bars.

When Lewis was five, he found a lizard on his driveway. The lizard was sluggish and injured, having lost its tail in a bird attack. No one saw the attack, but that was Lewis's theory, and I'm sure he was right because he was in biologist mode that day. Anyway, Lewis watched over the lizard for a couple of days, using his knight skills to keep guard as its energy and normal coloring[33] returned. Eventually the lizard crawled back to wherever it had come from.

Lewis's lizard was resilient. Being attacked by a massive flying object was surely an unexpected disturbance for it—and so was being befriended by a five-year-old biologist/knight. But the lizard minimized damage by sacrificing its tail to spare its life. Then it was able to bounce back and regenerate its tail.[34]

In soccer, team tactics can enhance resilience; a team used to high-scoring games is more likely to recover from an unexpected goal than a team built for defensive battles. Fans can boost resilience too, as was the case at halftime in Istanbul when the resilient Liverpool fans responded to a seemingly impossible three-goal deficit by chanting "We will win 4–3."

Diverse systems with multiple options are less vulnerable to external shocks and are therefore more resilient. Or, in lizard terms, the ability to temporarily function without a tail and then regenerate a new one made the initial bird attack more tolerable. Resilience is not the same as rigidity. If the lizard had insisted on keeping its tail, the bird would have eaten the rest of the lizard's body too.

In soccer terms, players with diverse skill sets contribute to resilience. Falling behind by a goal may require a change in tactics—and with sub-

33. This was one of those lizards that can change color between brown and green, depending on which shade best disguises it in its surrounding environment. I think the lizard found it difficult to match the color of Lewis's palm.

34. I didn't actually see the tail regenerate, but Lewis promised me it would, and my Internet research confirmed that the type of lizard he found is, in fact, able to regenerate a tail.

4

stitution limits, it's better to have players who can make these tactical changes on the fly. Leaders on the field add resilience too, by instilling changes without detailed instructions from the coach.

We will encounter bad bounces and unlucky losses that are beyond our control. But we can plan for longer-term resilience, and it's resilience that makes soccer memories. Spain's eventual triumph in the 2010 World Cup was even sweeter because they had lost their first game in that tournament.

We try to anticipate but can never know exactly what will happen. It could be an attacking bird, an unexpected goal, a volcano that disrupts travel plans, or a truncated warm-up. Resilience is about whether this unexpected event breaks our system or whether we minimize damage and come back stronger.

Resilience is resisting damage from an unexpected disturbance and then recovering quickly. The team that overcomes an undeserved deficit is resilient, and so is the lizard that got a little help from my five-year-old neighbor.

ZIDANE AND DISAPPEARING ICE

Threshold Crossing

It was one of those moments—I'll never forget where I was and how I felt. I was watching from State College, Pennsylvania, in the living room of our townhome, which was as bare as you would expect when I tell you that my Ph.D. student income was less than my professional soccer salary. I was alone because there are very few people with whom I can watch important games.[35] One such person, Monica, had decided she'd take a nap and just wait to see the next World Cup in four years. So there I was, sitting on a stool, and staring in disbelief at the repurposed computer monitor when my soccer idol,[36] the French midfield genius Zinedine Zidane, was ejected during overtime of the 2006 World Cup final.

Zidane is one of the best players ever and my favorite, so I was going to fit him into this book no matter what. Fortunately, there is no better example of a soccer threshold than the tragedy of Zidane in the 2006 World Cup final.

Zidane's last act of that game, and of his playing career, was an extra-time head-butt to the chest of the Italian defender who had been annoying Zidane (and me) all game. Zidane received a red card for the head-butt, which meant that he got ejected from the game and France had to continue with one less player. With Zidane ejected, Italy went on to win the game in penalty kicks.

It is not a stretch to reason that France would have won had Zidane not been ejected. He could have scored or created a winning goal in the remainder of the overtime. Even if that didn't happen, Zidane would certainly have been one of France's five shooters when the game progressed to penalty kicks.

But we'll never know what would have happened had Zidane stayed in the game. Red-card ejections are an accepted risk for reckless soccer players. Zidane often tested the line separating brilliance from bedlam,[37] a playing style that amplified his magnetic appeal.[38]

Zidane's story in that 2006 World Cup final is so compelling because of the contrast between his rash head-butt and his unflappable behavior earlier in the game. On his first-half penalty kick, Zidane used the Pan-

35. "Important games" looks like an oxymoron when I write it. So, clearly, this is a case where words are misleading.

36. Zidane's performance in the 1998 World Cup inspired my goal celebrations, my choice of shoes, and a red card that would have ended my college career if not for some fortunate results in games between other teams.

37. Zidane was also ejected from a game in the 1998 World Cup. He came back later in that tournament and scored two goals in the final, which France won. The head-butt in the 2006 final wasn't Zidane's first. He was ejected for that same offense in the 2000 Champions League while playing for the Italian club Juventus.

38. To get a sense of Zidane's appeal, I recommend the artistic documentary *Zidane: A 21st Century Portrait,* filmed during a single game while he was playing for Spanish club Real Madrid. Spoiler alert: he got ejected from that game too.

enka technique to put France ahead 1–0. If you don't know what that technique is yet, don't worry, half of an upcoming section is devoted to Antonín Panenka's gift to the Game. For now, just trust me that pulling off a Panenka in a World Cup final is as cool as it gets for a soccer player.

In that fateful 2006 final, Zidane passed a threshold, moving from one stable state (playing in the World Cup final) to another (ejected from the World Cup final). What's more, Zidane passed the threshold abruptly, which means that the resulting change (Zidane's ejection) was far bigger than the event that caused it (the Italian defender's verbal provocations).

Thresholds are points at which a relatively small change in one part of a system will cause a disproportionate change in the overall system. Once a threshold is crossed, the system may never return to the original state.

It's not just fans of volatile players who worry about thresholds. Abrupt crossing of thresholds is an urgent concern among scientists who study climate change, and among those who take scientists' findings seriously.

The climate system is nonlinear, which means that effects are not necessarily proportional to causes, which means that seemingly small tweaks can produce huge responses. Melting ice sheets in Greenland are one of the many[39] nonlinear events that experts believe could trigger abrupt and irreversible passing of climate thresholds—or may already have.

Images of lonely polar bears floating on dwindling chunks of ice tug at heart strings, but we'd be fortunate if stranded polar bears were our biggest concern related to melting ice. The ice sheets that are melting are massive, big enough to hold all the world's polar bears and the Coca-Cola that, according to television commercials, they enjoy drinking.[40]

39. Another worrying climate-change threshold is the acidification of oceans as a result of the excess carbon dioxide in the air. Higher acidity dissolves carbonate-forming organisms like coral. As these organisms dissolve, the ocean could pass a threshold from absorbing carbon to producing it, from being part of the climate-change solution to being part of the problem.

40. The Greenland ice sheet is 656,000 square miles, and shrinking. There are about twenty thousand polar bears left, and shrinking. Roughly 1.7 billion servings of Coca-Cola products are consumed each day, and let's assume that there is a year's supply on hand. If my math is correct, that means each polar bear could have thirty-three square miles of ice—they'd just need to share it with about thirty million servings of Coca-Cola products. So, yeah, it would all fit.

What's more, because white ice mostly reflects sunlight and the dark ocean mostly absorbs it, the melting ice sheets feed a reinforcing feedback loop: less ice leads to more absorption of solar heat, which leads to less ice, and so on.

The melting sheets contain enough water to raise sea levels to where low-lying cities such as Beijing and New York (and Maputo) would become uninhabitable. And the influx of cold freshwater from the melting might divert the Gulf Stream waters that now flow past England and Norway, which could[41] send Western Europe into an ice age.

Abrupt changes in the climate system are especially devastating when humans and the systems we depend on lack time to adapt. Relocating New York City residents and building protective structures around the city are costly measures, especially when we include social costs with the economic ones. But just reacting to events is costly too. In 2012, Hurricane Sandy tagged New York City and the surrounding region with roughly $70 billion in economic damage.

For Western Europe, the concern about sea-level rise is compounded by the diverted Gulf Stream waters that could make for a cooler region. Were the cooling to happen faster than natural temperature fluctuations, food crops would not adapt to the shorter growing season. And (remembering my delayed epiphany from the first chapter) disruptions to food supplies lead to negative health impacts, which lead to social unrest, and so on.

When average temperatures rise or fall slightly and over long periods, humans adapt and so do the systems that support human life. But we are testing adaptation limits with crossed climate thresholds that fuel extreme weather, water scarcity, flooding, and related health problems all over the world.

Like our Earth's climate system, Zidane adapted for most of the game. He adapted to constant shadowing from the noble Italian defenders. He adapted to an uncalled foul targeting his injured shoulder courtesy of the obnoxious defender he later head-butted. He adapted to

41. I'm using lots of "mights" and "coulds" in this section because the climate system is unpredictable. But unpredictability is not an excuse for ignorance where the "precautionary principle" applies. This principle says that even when we don't know for sure that something will happen, if the results of that something happening would be irreversible, we should do all we can to reduce the chance it will happen—like we would for a heart attack. Or put another way, when climate models tell us there is a 50 percent chance of passing a threshold and our only response is to hope for the best, that is like consenting to a round of Russian roulette when three out of the six revolver chambers have a bullet.

physical exertion and to a range of emotions, from elation after his successful Panenka, to disappointment when Italy tied the game, to frustration when his late-game header was saved by the Italian goalkeeper. Ultimately, it was a tasteless comment about Zidane's sister[42] from the soon-to-be-head-butted Italian defender that pushed Zidane past the threshold.

Like the climate system, Zidane could brush off a few fouls. The tragedy in his crossed threshold was that instead of hoisting the World Cup trophy in his final game, he walked past it with his head down.

Because of thresholds, even a small change can destabilize a system. Thresholds prevented Zidane from winning a second World Cup and are why I worry about climate change harming me (not to mention any future grandkids).

42. According to an Italian lip-reader, the defender said, "I wish an ugly death to you and all your family." Zidane's sick mother had been taken to the hospital earlier that day. Lip-readers from Brazilian television claim the defender called Zidane's sister a "prostitute." The head-butted defender even admitted, "Yes, I was tugging his shirt, but when [Zidane] said to me scornfully 'If you want my shirt so much I'll give it to you afterwards' . . . I answered that I'd prefer his sister, it's true."

Review: Behaviors

Every part of a system is connected, which means that each part's behavior is shaped by relationships to the other parts of the system. So, we can study elements, stocks, flows, feedback loops, and purposes, but we cannot know how the system works without also seeing interdependencies.

In a self-organizing system, behavior comes from a few shared initial conditions and from interactions among the parts of a system.

Emergence is when entirely new behaviors arise in a system, new behaviors that were not present in any of the parts alone. Therefore, emergent behaviors depend on system relationships.

Resilience is the ability to resist damage from an unexpected disturbance and then quickly recover.

Thresholds are points at which a relatively small change in one part of a system will cause a disproportionate change in the overall system. Once a threshold is crossed, the system may never return to the original state.

5

Evaluating

To evaluate sustainability in our systems, and check our progress, there are properties to consider and methods to apply.

MOZAMBIQUE'S CIVIL WAR AND PORTUGAL'S BEST WORLD CUP

Overview: Evaluating Sustainability

The Sacred Monster's first and only World Cup looked to be finished as Portugal fell behind 3–0, just twenty-two minutes into their 1966 quarterfinal matchup with a surprising North Korean team.[1] But Portugal scored once and then didn't stop. By the time the game ended, Portugal were comfortable 5–3 winners and headed on to the semifinals.

Portugal changed that game's *inertia,* which is a resistance to change in the state of motion. When we recognize inertia in a system, we see ways to change the state of motion or even use it to our advantage, instead of hopelessly fighting it. Inertia is why systems at rest tend to stay at rest and why systems in motion tend to stay in that same state of motion. And inertia is why the first goal in a comeback is so crucial. After scoring their first goal, Portugal were still behind 3–1, but they had reversed the game's inertia.

Advancing to the semifinal game, Portugal came up against England, who were also the World Cup host. For Portugal, it was an unlucky instance of *path dependence,* in which present circumstances are limited by choices from the past, even when past circumstances are no longer relevant. Or, in

1. Making their first World Cup appearance, that 1966 North Korean team was the first from Asia to reach the quarterfinals. North Korea's team has qualified for only one World Cup since then.

soccer terms, Portugal faced a huge disadvantage in that they were playing against the home team[2] in that semifinal (because England had been selected as World Cup host years before), even though Portugal had outperformed England in previous games at that World Cup.[3]

Facing an uphill battle as the away team,[4] Portugal lost. England went on to win the World Cup.

The Sacred Monster was also in the prime of his career four years earlier, in 1962, and so was Eusébio, the Black Panther. So Portugal must also have had a chance at the World Cup that year, right?

That depends on how we define *chance*. Portugal did have the chance to qualify for the 1962 World Cup, just not to play in it.

During qualification, Portugal and England were assigned to the same four-team group. England won all their games to earn the group's single allocated spot in the 1962 World Cup in Chile. At the time,[5] sixteen teams was the World Cup's *carrying capacity,* which is the population that an environment can sustain indefinitely, given the available resources. So Portugal missed out, and the world had to wait four more years to see the Sacred Monster and Black Panther on the Game's biggest stage.

Missing a World Cup is trivial compared with how carrying capacities can shape off-the-field events. During Mozambique's postindependence civil war, the combatants were seemingly fighting for power. But the political conflict was aggravated by droughts that shrank the region's carrying capacity for human life, which led to fighting over basic needs like water and food.

Path dependence, inertia, and carrying capacities are properties to consider as we evaluate systems. Considering these properties sheds light on the Sacred Monster's quest for the World Cup—and on our quest for sustainability.

2. Playing at home is a huge advantage. Teams are roughly twice as likely to win a home game as they are to lose it, according to intuition and to *Statistical Thinking in Sports* by Albert and Koning.

3. Prior to their meeting in the semifinals, Portugal had won all four of their games at that World Cup while England had won three and tied one.

4. Not only did Portugal have to play that 1966 World Cup semifinal game in England, but the location was changed from Goodison Park in Liverpool, which held around sixty thousand fans, to Wembley Stadium in London, which could pack in close to a hundred thousand. The change, made to increase ticket revenue, also increased England's home-field advantage.

5. The World Cup tournament was expanded to twenty-four teams in 1982, and to thirty-two in 1998. FIFA is constantly discussing adding more, so don't be surprised if it's forty or forty-eight for Qatar 2022.

In addition to considering specific system properties, we also need to evaluate sustainability more generally. Methods to help us do so include counterfactuals, the five whys, and life-cycle analysis.

Counterfactuals are a systematic way of asking ourselves "what if" questions in order to consider alternative outcomes. In the case of Mozambique's civil war, it's tempting to blame the warring groups, which might lead us to assume that getting rid of these groups would stop the war. Counterfactual thinking forces us to contemplate what could happen if the groups were in fact disbanded. Then we realize that a probable outcome would be different groups in the same fight for postindependence power and drought-ravaged resources.

After using counterfactuals to consider alternative outcomes, we can seek root causes of unsustainable behavior using *the five whys,* a method of repeatedly asking why in order to uncover new details. Asking why about Mozambique's civil war tells us about the power vacuum. Asking why again tells us that the huge death toll was due to famine made worse by war, and vice versa. Asking why one more time tells us that the ruling party's war effort was propped up by the United States, Russia, and France and that the resistance movement had support from Rhodesia and South Africa. As we learn more about the real causes of unsustainable behavior, we are better prepared to fix them.

We can measure these causes and their effects using *life-cycle analysis,* a method to discover the value added by a product or service as well as the resulting sustainability impacts over a period of time. Mozambique's civil war lasted from 1977 until 1992 and caused nearly a million deaths. That simple example of life-cycle thinking is enough to confirm that, even if there were some sort of value added by the civil war, it could never offset the war's terrible costs.

On a happier note, when Portugal beat the defending champions Brazil in the opening round of that 1966 World Cup, a stunning goal by Eusébio sealed the game. People all over the world watched the goal in televised highlights. And those of us who weren't watching in 1966 can easily find and watch the goal on Youtube. The Black Panther's legend is secure.

In other key moments from that game, the Sacred Monster ran to defensive positions that stopped Brazil's attack and prevented goal-scoring opportunities—all without ever touching the ball. But highlights don't show players running to smart defensive positions.

We don't intentionally ignore the Sacred Monster, and we don't set out to create sustainability problems; but both occur when we fail to consider effects that seem too small to matter or too distant in time or space. We can avoid such oversights by evaluating sustainability with methods like counterfactuals, the five whys, and life-cycle assessment.

LAMPARD AND ME, HIGHWAYS AND RAILROADS

Path Dependence

England midfielder Frank Lampard and I were born less than a month apart in the summer of 1978. Lampard was born in London to a father who made his living as a professional soccer player.[6] I was born in the northeastern United States to a father who made his living as a biology professor.

In 1994, Lampard joined the youth team of West Ham United, the professional club for which his father had played and was then an assistant coach. The young Lampard played with teammates on track for professional careers and was immersed in soccer; he attended weekly professional games and followed the ever-present coverage in newspapers and on television. At West Ham, Lampard got to play with his hero, striker Frank McAvennie, whose professional career spanned fifteen years and included stints with the Scottish national team.

In 1994, I joined the best amateur club team around, the Syracuse Blitz, and got to play with teammates on track for college scholarships and college acceptances made possible by soccer. I learned about the Game on a one-week delay from *Soccer America,* a newsmagazine that covered professional leagues, international tournaments, and the U.S. college season (including a weekly player spotlight). I watched soccer on television when there was a World Cup, or when I put in the *500 Greatest Goals* video I got for my birthday. At seventeen, the best player I had ever stepped on the field with was a Greek guy who owned a local restaurant and had played a couple of seasons for a professional indoor team.

In August 1996, Lampard made his first start for West Ham's first team. In 2001, he left West Ham for Chelsea for a transfer fee[7] worth over $18 million (this was before Chelsea was bought by Abramovich,

6. Lampard's uncle is Harry Redknapp, another former player and successful coach who once allowed a fan to play in a preseason game. Don't worry, there's more to come on that story.

7. Transfer fees are money paid from one club to another for the rights to a player. These fees are a rough projection of the player's value for the rest of their career. Lampard has earned well over $18 million since the transfer from West Ham to Chelsea.

the Russian oilgarch). Lampard has since scored over two hundred professional goals, appeared over one hundred times for England's national team, and dated supermodels.

In August 1996, I enrolled at Lafayette College and made my first start at the highest level of college soccer. What followed were the most rewarding four years of my soccer life. I was even featured in that weekly player spotlight in *Soccer America,* which I think probably had as much to do with my dad being one of the few long-term subscribers as it did with my on-the-field achievements. I graduated from Lafayette at the same time as teammates and friends who have gone on to careers as teachers, doctors, lawyers, clothing designers, and investors. After I graduated, I played in a new professional outdoor league and in the same indoor league the Greek restaurant owner had played in. In 2001, I retired from my low-paying professional soccer career; by 2008, I was working as a professor.

Is it any surprise that our respective careers turned out the way they did?

Of course, our birth conditions don't determine everything. But we also need to ask ourselves, "What path are we on if we don't do anything new?" The answer was different for Lampard than it was for me, and the answer is different for a child in a poor family than it is for a child in a wealthy one.

Path dependence is when small initial differences in a system are a disproportionate cause of differences in outcomes. Or, in other words, history shapes the present.

Path dependence doesn't just influence soccer career paths; path dependence also affects development patterns, which in turn shape sustainability.

Funding for the U.S. interstate highway system was authorized in the 1950s, mostly so that troops and military equipment could move rapidly around the country. Today, the huge subsidized network of roads makes it tough for trains and other forms of public transportation to compete economically.[8] When a highway already connects two destinations, there is less motivation to invest in a railway covering the same

8. Still, investments in public transit can pay off. In their *Transportation Funding and Job Creation* report, Smart Growth America found that stimulus investments in public transit created about a third more jobs per dollar than investment in new road and bridge construction. So, compared with relentless new building, public transit can be a better investment for both the economy and the environment on which it depends.

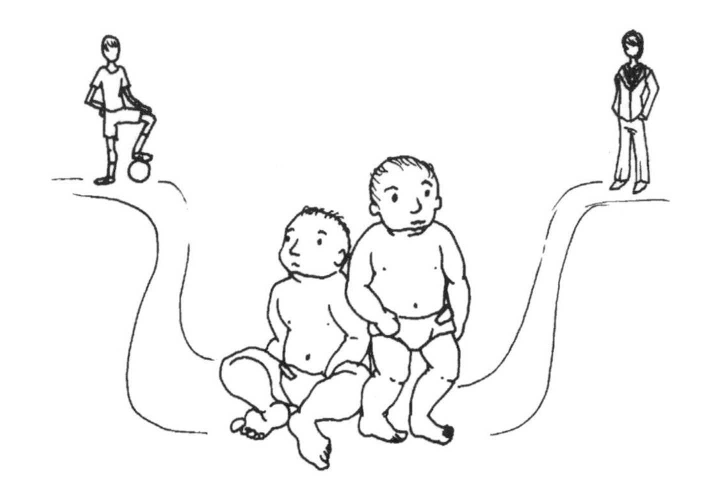

path, even if that railway could potentially be more convenient, better for the environment, and cost less on a per trip basis.

But by subsidizing highways we subsidize sprawl, long driving commutes, and disconnected neighborhoods. By subsidizing highways we encourage the travel that uses fossil fuels, and so we subsidize the climate-changing emissions we desperately need to curb.

Don't get me wrong, I appreciate the interstate highway system; it's reassuring to know I can cover the thirteen hundred kilometers to see my parents in twelve hours.[9] But now, more than half a century after the interstate highway system was authorized, we remain stuck on the path made by that decision.

So there's no avoiding existing path dependence as we strive for more sustainable systems, but we can recognize it and act accordingly. We can also be proactive, considering the implications of paths we establish with our present-day actions.

I'm not saying there was no chance for me to have Lampard's career; I could have been more thoughtful in my practice and sought out more challenging competition. I'm just saying that path dependence made it so that the chance I would play for West Ham and Chelsea was about the same as the chance of Lampard becoming a professor.

9. I do drive a Prius on long road trips, but I would get better gas mileage if I drove it slower.

Now, I wouldn't trade places and I don't imagine Lampard would either.

Path dependence is when our present-day choice options are shaped by past decisions, even if past circumstances are no longer relevant. Path dependence makes it difficult for public transportation to compete economically in the United States and for me to compete economically with the English midfielder Frank Lampard.

PANENKA'S GIFT AND NEW JERSEY DUNE GRASS

Inertia

One of my biggest soccer regrets is not learning about Antonín Panenka's gift to the Game until my playing days were over. I absolutely would have mimicked his innovation at some point. Unlike the genius Panenka, however, I would never have had the audacity (or opportunity) to try it with such high stakes.

Probably the least famous of the famous soccer players mentioned in this book, Panenka deserves to be a legend. He was a creative attacking midfielder for some of the best professional club teams in Europe and for the Czechoslovakian[10] national team. Panenka was even named Czechoslovakian player of the year in 1980. By that time, he had already made his most enduring contribution to the Game—and to sustainability analogies.

In the final of the 1976 European Championship, Czechoslovakia and West Germany were tied 2–2 after regulation and overtime, so the game was decided by penalty kicks. The first seven shooters (four for Czechoslovakia and three for West Germany) all scored. Then the fourth West German shooter missed. Panenka stepped up to the penalty spot with the chance to win the game and the tournament.

Millions of players all over the world can only dream of the opportunity Panenka had—just him and the goalkeeper; score the twelve-yard shot and his team are champions, and he is a national hero.[11] Yes, Panenka scored, but it's the manner in which he did so that is legendary.

10. Czechoslovakia was a nation from 1918 until 1993, when it was peacefully divided into the Czech Republic and Slovakia. In an early indicator of the coming official division, the Czechoslovakian national team sometimes held two separate practices—one for Slovakian players and one for Czechs.

11. In Panenka's case, he was hero to a nation that would cease to exist after 1993.

On his hero-making shot, Panenka kicked the ball slowly, waist high, and straight at the center of the goal. Or, put another way, Panenka shot the ball right where the West German goalkeeper, Sepp Maier, was lined up.[12] If Maier had stood still, he would have easily caught Panenka's tame shot.

But Panenka guessed that Maier would dive early to one side or the other—trying to anticipate a shot to one of the corners of the goal. Panenka's hunch was correct; Maier dove to the left, and his momentum carried him out of position as Panenka's gentle shot floated into the recently vacated center of the goal.

Shooters have the upper hand on penalty kicks. Even the best goalkeepers can't save a well-placed and hard shot by simply reacting. So, goalkeepers will try to improve their chances by guessing a side and moving in that direction just as the ball is kicked, or even before.[13]

When the goalkeeper guesses the correct side of the goal, even a perfect shot can be saved. Some shooters respond by attempting faster and more precise shots to the corner, but shots like these have a higher probability of missing the goal altogether.

Panenka is a genius[14] because he accepted that the goalkeepers were moving early and, instead of resisting the movement, looked for the new opportunities that the movement presented. Panenka figured out that, when a goalkeeper moves early, a slow shot down the middle works, no matter which direction the goalkeeper guesses. Panenka's penalty took advantage of the goalkeeper's inertia.

Inertia is a resistance to change in the state of motion—no different from the inertia concept I had forgotten from my first physics course. Inertia is why systems at rest tend to stay at rest and why systems in motion tend to stay in that same state of motion. When we know the state of motion, we can use it to our advantage instead of hopelessly fighting it.

12. The video of Panenka's penalty is on Youtube and a little grainy, but it's definitely worth watching.

13. There have been periods when it was illegal for goalkeepers to move early, but they did it anyway and it was rarely called (yes, I'm still annoyed that illegal early movement led to a couple of my penalties being saved).

14. The Brazilian soccer legend (and national treasure) Pelé said that anyone who took Panenka's approach to penalties was "either a genius or a madman." I contend that Panenka deserves recognition as the former.

My favorite non-soccer[15] inertia examples are from my summer vacations. I go with my family to Ocean City, a barrier island in southern New Jersey that manages to be both kid-friendly, with rules prohibiting alcohol sales on the island, and adult-friendly thanks to the massive liquor stores gracing the bridge approaches onto the island.

Sandy barrier islands like Ocean City are dynamic environments that, left alone, would naturally migrate up and down the coast as the inertia in ocean currents pushes the sand around. Barrier islands are susceptible to hurricanes and flooding; it is not uncommon for flood water to entirely overtake Ocean City, connecting the bay to the ocean.

Despite the naturally inhospitable environment over time, people like me still love the beach. The cottage we visit each summer has been in our extended family for over a century (with many of the original amenities). It's a connection to my nieces, nephew, cousins, aunts, uncles, and generations of people I never got to meet and who never had to meet me. My wife, Monica, and I were married[16] within walking distance of the cottage.

Truly respecting the inertia in the ocean currents would mean not building houses right next to the ocean, but when it comes to Ocean City, I need a middle-ground solution between abandoning the island altogether and living on it more sustainably.

Every quarter-mile or so on the Ocean City beach are jetties, which are stacked rows of jagged black boulders extending perpendicularly from the beach out into the ocean. The jetties of Ocean City, and other barrier islands, are intended to reduce storm impacts, keep beaches intact, and prevent the islands' natural movement up and down the coast.

Sometimes jetties are a suitable option, especially for those who get paid to design, build, and maintain them.

But jetties can also accelerate the loss of beach sand, which is pretty much the opposite of what they are intended to do. Waves that come up against a gradual beach slope are like a ball that rolls up a gradual slope and then back down. Waves that smash against jetties are like a rolling

15. Here's a bonus soccer example of inertia: in addition to laziness, inertia is one reason why some of the best goal-scorers seem to loaf around (see Linekar, Gary; Henry, Thierry; and Berbatov, Dimitar). They lull defenders into a false sense of security and then spring into action to create a decisive goal.

16. Getting married in a World Cup year ensures that I'll never forget how many years I've been married. June 10, 2006, is forever etched in my memory, partly because, between the ceremony and reception, I watched Argentina play Ivory Coast in one of the most anticipated first-round games of that World Cup.

ball that bounces off a wall. Because the wave retreats faster, more sand is carried back into the sea with the retreating water.

By building jetties, we engage in an unwinnable arms race against the ocean's inertia. On barrier islands, which are nothing but sand, trying to control Nature with jetties doesn't work for very long.

Compared with artificial jetties, sand dunes offer better protection from storms and waves and are less expensive and more beautiful. Sand dunes are basically "biiiiig piles" (my three-year-old nephew's perfect description) of sand between the ocean and whatever it is we want to protect, like our family vacation house. But sand alone will eventually blow or wash away.

Dunes can last, and even grow on their own, when there is dune grass and sand fences. Extending deep into the sand, the roots of the dune grass control natural dune erosion. Sand fences slow down wind, another form of inertia in the barrier island system. When the wind slows down around the fences, this causes sand to drop out of the wind and settle next to the fences. As more sand accumulates, the growing pile slows wind, which causes even more sand to drop out.

Both the grass and the fences use the inertia of sand and wave movement already in the barrier island system to their advantage, rather than trying to resist this inertia like the less effective but more expensive jetties. By intentionally placing dune grass and sand fences, we can encourage sand accumulation where we want it.[17]

It takes courage to trust our interpretation of the inertia in a system. A conventional well-placed penalty that is saved by a guessing goalkeeper gives the shooter an excuse. A saved slow shot down the middle of the goal makes the shooter look like an idiot.[18]

Heroic players, like Panenka in 1976 (and Zidane thirty years later), choose the approach they think gives them the best chance to score. They ignore the real but irrelevant risk of personal embarrassment.

In the same way, courageous designers and elected officials will reject the visible and status quo jetties that best protect themselves from criticism in favor of the dune grass and snow fences that best protect their constituents from flooding and lost beaches.

17. Sand fences also keep off the dunes those who don't read or obey signs that say "Please keep off the dunes."

18. Youtube can show you that Neymar, Robin van Persie, Andrea Pirlo, and Alexandre Pato are among the stars who have suffered the embarrassment of unsuccessful Panenka attempts.

2

Even more than courage, using the inertia of a system requires study and practice. Conceptual designs, laboratory tests, and field experiments tell us that dune grass and sand fences protect against beach erosion. Panenka observed and reflected on the penalty-kick system and then relentlessly worked on his new technique after other players had left practice. So, when it came time to debut his gift on the biggest stage, Panenka was "one thousand percent certain" (his words) that he would score.

Inertia is a resistance to change in the current state of motion, which means that, by using inertia, we can avoid wasted effort. Because of inertia, simple dune grass controls beach erosion, and slow penalty kicks fool guessing goalkeepers.

ARGENTINE DEFENDERS AND UNSUICIDAL LEMMINGS

Carrying Capacities

Javier Zanetti earned his perfectly descriptive soccer nickname, "The Tractor." From his typical position as a right defender, Zanetti ran perpetually up and down the field, plowing unfortunate forwards who got in his way.

Zanetti played for Argentina more times than any other player, representing his country in 145 games between 1994 and 2011. His club career was also exceptional; the 845 games he played for the Italian club Inter Milan is another record.

Toward the end of his career, Zanetti captained Inter to their finest season as they won both the Italian league and the European Champions League in 2010. In the Champions League semifinals, Inter upset the defending champions FC Barcelona because Zanetti anchored Inter's stifling defense (and, as we covered earlier, because Barcelona were tired from their volcano-induced bus trip). Then, in the final, Zanetti's Inter shut out FC Bayern Munich, a team loaded with players from Germany's national team. Zanetti triumphantly raised the Champions League trophy on May 22, 2010.

Less than a month later, Zanetti's Argentina were progressing through the 2010 World Cup. Yet even as they won their first four games to advance to the quarterfinals, Argentina seemed vulnerable.

Sure enough, when Argentina faced the always-strong Germans[19] in the quarterfinals, it was a rout. Argentina conceded a goal just three minutes into the game and were hammered out of the tournament by a score of 4–0. All four of Germany's goals exploited open space on the right side of Argentina's defense.

What happened to the great Zanetti?

He wasn't on the team.

I'm with you if you think it may have helped Argentina to have one of the best defenders ever, coming off one of the best seasons ever, play against a team full of players he had beaten less than a month before. But the person whose opinion mattered disagrees with us.

Argentina's manager for the 2010 World Cup was the legendary ex-player Diego Maradona. Fans voted Maradona the best player of the twentieth century,[20] and he is a godlike figure[21] to many people in Argen-

19. The English player Gary Linekar was only slightly exaggerating Germany's soccer track record when he said, "Football is a simple game. Twenty-two men chase a ball for ninety minutes and at the end, the Germans always win."

20. Maradona was named player of the century by fans voting in an online poll (journalists and coaches selected Pelé). He rose from staggering poverty and led Argentina to victory in the 1986 World Cup. In the semifinal against England, he scored two of soccer's most famous goals: one with his hand (the referee didn't see it) and one by dribbling past half of the English team. Later in his career, Maradona's cocaine habit led to a year-long suspension; after he returned from that suspension, he tested positive for ephedrine and was kicked out of the 1994 World Cup. Maradona remains one of the world's most fascinating personalities, soccer or otherwise. He fired an air rifle at reporters, had an audience with two different popes, befriended Fidel Castro and Hugo Chávez, and was (falsely) reported dead three separate times in 2007.

21. There is even a "Church of Maradona," whose followers count the years since his birth in 1960 and follow commandments such as "Name your first son Diego."

tina, around Latin America, and in Italy, where he lifted his Napoli club to international prominence.

The reverence left over from Maradona's playing days contributed to his ill-fated selection to lead Argentina's national team. Sure, ex-players can become winning managers, but it's typically a good idea to have them prove themselves as managers before handing them a national team. When Maradona took the reins of Argentina, he hadn't managed in over a decade, following a couple of short and unsuccessful stints with club teams.

Regardless of their qualifications, managers have final say in the players who will go to the World Cup (sometimes politicians give "advice," as when the Cameroonian president picked the striker who went on to embarrass El Loco Higuita). World Cup rosters are limited to twenty-three players and cannot be added to once the tournament begins.

Even before Argentina crashed out of the 2010 World Cup, fans and media ridiculed Maradona[22] for not selecting Zanetti. The Tractor wasn't going to the World Cup, but Ariel Garcé was, and Garcé had played for Argentina just twice in the previous five years.[23] Martín Palermo was selected even though he was a washed-up thirty-six-year-old who was totally redundant as the sixth-best forward on a team that would only play two at a time. Zanetti watched the 2010 World Cup like the rest of us, as his position on the right side of the defense was patrolled by either Nicholás Otamendi, who was not used to playing that position, or by Jonás Gutiérrez, who played his club soccer for an English team that wasn't even in the top division.

Maradona's explanation for his inexplicable roster was "I have no doubt about this list."

So Zanetti missed the 2010 World Cup because, after including the worthy players and whims of Maradona, there was no carrying capacity remaining on Argentina's twenty-three-player roster.[24]

22. Maradona responded to Pelé's criticism with "I would tell Pelé to go back to the museum." He shrugged off the French legend Michel Platini with "We all know what the French are like and Platini is French. He thinks he is better than the rest. I pay no attention." To the media in general Maradona said, "I don't care what you people think about me as a manager. I couldn't care less. I've got my twenty-three players and I'm ready to die with them."

23. Rumor has it that Garcé was chosen because Maradona had a dream in which Argentina won the tournament and Garcé played a key role.

24. Maradona resigned after the 2010 World Cup. By August of that year, with a new manager, Zanetti was right back playing right back for Argentina.

Carrying capacity is the population that an environment can sustain indefinitely, given the available resources and the rate at which they are being consumed.

Carrying capacity explains why Zanetti didn't play in the 2010 World Cup, and carrying capacity clarifies the seemingly suicidal behavior of lemmings. It's a popular myth that these small furry rodents willingly follow each other and run off of seaside cliffs to their certain death in the waters below. Allegedly, the lemmings behave this way to intentionally curb overpopulation.

In reality, lemmings are not conformist or suicidal. Those are misconceptions that were perpetuated by misleading "scientific" documentaries such as *White Wilderness,* for which filmmakers created the scene by using deceptive camera work—and by launching helpless lemmings from a cliff.

Sure, a few lemmings may happen to jump from cliffs, but not because they have been brainwashed into committing mass suicide. Any cliff-jumping lemmings are making a calculated attempt at migration. Lemmings can swim about two hundred meters in calm water and will do so to find a new habitat.

Even though they can swim, lemmings will kill each other competing for scarce local resources before they resort to jumping from cliffs and swimming to seek resources elsewhere. It's kind of like how soccer players will compete for a starting position on their current team before looking to move to another team with an open spot.

Whether a lemming fights or jumps, it is responding to competition for food, space, or mates, all of which are signs of exceeded carrying capacity.

So, despite the popular myth, lemmings are no more suicidal than Spanish explorers, Irish immigrants, or Haitian refugees crossing oceans to seek better lives in other lands. Lemmings are no more suicidal than Zanetti was when he traveled from Argentina to Italy to find his place in professional club soccer. Humans, like lemmings, seek environments with the resources to sustain life—environments that have available carrying capacity.

Rural Syrian farmers seek food and shelter when they move into cities after severe droughts force them to abandon their crops. Cities have more resources, but they also have more people who need them, and more people who are using more than they need. At some point, we run out of places with unused carrying capacity to migrate to. Then we are forced to choose between starving and fighting our neighbors for food. In Syria this influx of new migrants has triggered a civil war in which hundreds of thousands of people have been killed, and those who try to flee risk ending their journey with drowned toddlers washing up on European beaches.[25]

In other words, we've got problems if the impact of humans exceeds the carrying capacity of Earth.[26] The I = PAT equation is a simplified but useful guide to measure human impact. The equation describes the environmental impact (I) of population (P) multiplied by affluence or level of consumption (A) multiplied by the technology or processes used to get resources and turn them into useful goods (T). The equation allows us to estimate, for example, that an environment that is currently at capacity supporting one million people could support two million if everyone consumed half as much (assuming that the technology variable was unchanged).

Again, IPAT is a simplified equation that doesn't give us exact answers. But it provides guidance on some of the core questions about carrying capacities and sustainability:

> Are our technological gains offsetting rising consumption and population? It doesn't look like it.
>
> Can we set limits on consumption? Perhaps for me, but certainly not for the billions of people living in poverty.
>
> Will we control population by empowering men and women with education and birth-control options? Or will we let exceeded carrying capacities do it for us? The first option sounds better to me.

25. Peer-reviewed scientific studies are now connecting climate change and war. One of the most comprehensive studies is described in this article: Kelley, C., Mohtadi, D., Cane, M., Seager, R., and Kushnir, Y. (2015), "Climate change in the Fertile Crescent and implications of the recent Syrian drought," *Proceedings of the National Academy of Sciences USA,* 112, 3241–3246.

26. By some estimates, we are already exceeding the carrying capacity of the Earth with our impacts. In a 2001 United Nations report, estimates for the human population that would stay within Earth's carrying capacity ranged from four billion to sixteen billion people, depending on assumptions for variables such as consumption and waste levels. The world population is now over seven billion people.

I'm not sure of the answers to these questions. What I am sure of is that ignoring them is dumber than letting Maradona choose a national team, and more suicidal than a swimming lemming.

Carrying capacity is how much population can be supported, whether that population is soccer players on a World Cup roster or lemmings on an island.

MY MISSED PENALTY AND A STERN REVIEW

Counterfactuals

My junior year in college, in our league championship game, my poorly taken penalty kick was saved by the goalkeeper. We were already winning 2–0, and there were only about twenty minutes left in the game, which we were dominating. We closed out the game pretty easily, avoiding needless anxiety for the player who had missed the penalty that could have extended our lead.

Sometimes I think back to that penalty kick and wonder what it would have been like to make it. Scoring a game-clinching goal is one of the best feelings in soccer (and life). A clinching goal that gave my team a three-goal cushion at the end of the game would have brought not only the incomparable bliss of scoring, but also the relief of knowing we would win.

Had I scored on that penalty, I like to believe I would have gone to celebrate with my friend who had been fouled to earn the penalty (and had passed to me for a goal earlier in the game). More likely, I would have reacted by running to the sidelines and jumping over a short fence into a crowd of intoxicated college students.

Can we know how events would have unfolded, in my celebration and beyond, had I made that kick?

Counterfactual analysis is used to study alternative outcomes. It's basically the same after-the-fact second guessing that we do all the time, except more methodologically sound.[27]

For systems-level questions about sustainability, it's impossible to run controlled laboratory experiments in which hypothesized causes are manipulated and all other variables are held constant. There is no experiment that can tell us how those Bronx neighborhoods would look today if the expressway had never been built through them.

The economist Robert Fogel used counterfactuals to study experiment-proof questions. In fact, he won the Nobel Prize for his pioneering application of counterfactuals to study railroads and economic growth in the United States.[28]

Counterfactuals become more than after-the-fact second guessing when they have clarity and consistency. We create clarity by defining the what-if (I make the penalty kick) and the possible alternative outcomes (the result of that game). We foster consistency by defining connecting principles (the relationship between my goal and the status of the game) and by ensuring that these principles make sense theoretically. For example, the counterfactual analysis of my penalty should

27. My argument that Barcelona would have won three straight Champions Leagues if not for the Icelandic volcano would never survive a rigorous counterfactual analysis.

28. I used Fogel's counterfactual approach to study my missed penalty and to earn my Ph.D. My counterfactual thinking even goes back to when I was a toddler in a car seat; when my mom suggested that I "take a nap and we'll be there when you wake up," I asked her, "Where will we be if I don't take a nap?"

reflect the level of confidence (well over 99 percent)[29] that being ahead 3–0 with twenty minutes to play will result in a win.

Nicholas Stern, a former chief economist of the World Bank, used counterfactual thinking in his oft-cited forecast of the global economic impacts of climate change. Stern's review concluded that avoiding the worst effects of climate change would require an annual investment of 1 percent of global gross domestic product.

Investing that much to avoid the worst effects of climate change would be like eliminating Australia's entire gross domestic product, every single year. It seems like an impossible investment, until we consider that Stern's review forecasted that, by taking no action, our annual costs would be between five and twenty times higher!

It's like we are my young neighbor Lewis's resilient lizard and our attacking bird is climate change. We have a choice. Act now and give up the tip of our tail. Or wait, and lose our whole tail—and perhaps a leg or two.

Counterfactuals get more complicated as more time and causal steps are included between the what-if and the possible alternative outcomes. This is unavoidable for long-duration sustainability questions, such as

29. Sure, Liverpool came back in Istanbul, and Portugal came from three goals behind to beat North Korea in 1966, but neither feat has been duplicated in the respective competitions. Plus, Liverpool and Portugal each had nearly three times as much time to come back as our opponents would have. Plus, Portugal had Eusébio.

how an Everglades canal built in the 1950s affects a Miami teenager today. Another limitation is that the results from counterfactuals are only as good as the data we use for assumptions, probabilities, and uncertainties. My penalty-kick counterfactual would produce misleading conclusions if I mistakenly thought a team behind 3–0 had a 50 percent chance of coming back to win.

Such limitations are partially neutralized when we use counterfactuals to compare alternative scenarios. For instance, Stern used probabilities to represent variables in the climate system, such as frequency of storms and amount of sea-level rise. Sure, there is uncertainty in these variables; we don't know how many cyclones will hit Mozambique next year, and we don't know whether the cyclones will smash into Maputo or unload their energy on a less populated area. When our counterfactual is a comparison, however, we just assume the same storms in both scenarios. And when we use the same assumptions in both scenarios, we reduce the effect of uncertainty on our findings.

To study the economic costs of climate change, Stern's review applied the same probabilities to both the do-nothing path and the take-action path. Uncertainties about sea-level rise and storm characteristics make it impossible to know the exact cost for each scenario. But because the same educated guesses underpin both scenarios, we can be confident in the review's main message: doing nothing will cost far more than taking action.

Counterfactual analysis helps us systematically consider alternative outcomes. Counterfactual thought let me stop regretting a missed penalty kick, and counterfactual thought tells us that acting to avoid the worst effects of climate change will be expensive, but not nearly as expensive as doing nothing.

BARBOSA, BIGODE, AND THE *CHOICE* TO EAT DIRT

The Five Whys

In mid-July 1950, a startling number of suicides and heart attacks was reported in the seaside Brazilian city of Rio de Janeiro. A famous radio broadcaster shot himself. Less famous people drank poison and jumped to their deaths. But no prophet had predicted the end of days, and no invading army threatened Rio's shores.

The reality was worse.

On July 16, 1950, Brazil lost in the World Cup final.[30] The game was played in Rio's newly built Maracaña Stadium, in front of about two hundred thousand people.[31]

I think the Brazilians who killed themselves because of that lost soccer game acted irrationally.[32] Still, as the Brazilian anthropologist Roberto DaMatta explains, the loss in 1950 was "perhaps the greatest tragedy in contemporary Brazilian history. Because it happened collectively and brought a united vision of the loss of a historic opportunity. Because it happened at the beginning of a decade in which Brazil was looking to assert itself as a nation with a great future."

Brazilians expected that the 1950 World Cup would validate their nation's potential. When their expectation wasn't met in the final, a few people, probably already suffering from psychological illness, were pushed past the ultimate threshold.[33]

Theories abound as to why Brazil lost that fateful final. One way to test our theories and uncover new ones is by using "the five whys." The approach is straightforward: ask a why question, answer that question, ask why again on the basis of that answer, and so on. Five is just a guideline for the number of iterations; sometimes fewer will suffice, and sometimes more are needed.

For the 1950 World Cup, my first question was *why* the Brazilian team lost. Their goalkeeper Moacir Barbosa is an easy scapegoat. He was beaten for the game-winning goal at his near post, which is something goalkeepers are trained to protect against.[34] Even though journalists named Barbosa the best goalkeeper of the tournament, he never lived down that goal against him in the final. As illogical as it may have been, Barbosa missed out on future opportunities to represent his country. Even more than four decades later, Brazil's soccer federation pre-

30. The "final" was actually the last group-stage game but was effectively a final because of prior results. If Brazil won or tied, they would be champions. If Uruguay won, they would be champions.

31. That's not a typo. Two hundred thousand fans attended a soccer game, in 1950! It remains the largest crowd ever gathered to watch a sporting event in an enclosed stadium.

32. I also think that an announcer killing himself because of a lost soccer game is no less rational than when Brazil's sexist dictator Getúlio Vargas killed himself because of lost political power (more on that later).

33. In their book *Soccernomics,* Simon Kuper and Stefan Szymanski report that, in most countries, fewer people kill themselves when their team is playing in the World Cup.

34. Watch replays of the goal on Youtube and you will see that the ball took an awkward bounce right before reaching Barbosa. It was not an easy save.

vented Barbosa from visiting with the national team as they prepared for the 1994 World Cup.[35]

My next question was *why* the shooter was open in the first place. Here we find another scapegoat, Brazil's defender João Ferreira, who was also known as Bigode—"Mustache." Not only was Mustache beaten by his opponent on the goal in question, but he also failed to disrupt the build-up to Uruguay's goal, which had tied the game just thirteen minutes earlier. Mustache never played again for Brazil's national team.[36]

My third question then became *why* the score was so close that two goals in thirteen minutes were enough to defeat mighty Brazil. One theory for why Brazil had not put the game out of reach is that the entire nation, players included, was overconfident. Perhaps this is the reason the huge Maracaña crowd went silent and the Brazilian players seemed shocked after the goal that tied the game.

Thus, my fourth question was *why* Brazil was so confident. For starters, Brazil had yet to lose in the tournament and they needed only a tie to win the World Cup. Plus, before the final, the Brazilian team had been moved into a hotel closer to downtown Rio and away from their rural training ground, where they had been able to focus on soccer. In Rio, players were surrounded by adoring fans and used by enterprising politicians who wanted to associate themselves with the team in order to advance their own agendas. In Rio, the day *before* the final, a newspaper showed the Brazilian players next to the headline "These Are the World Champions." In a pregame speech, Rio's mayor addressed the team as "you who I already salute as victors."

For me, the big revelation came after asking *why* Brazil was *over*confident—the fifth question. It's simply confidence if you are 99 percent sure you will win and you actually have a 99 percent chance of winning. Overconfidence is when there is a gap between expectations and reality. When I asked why Brazil was *over*confident, I was forced to consider, for the first time, the team that beat them: Uruguay. It's easy to assign blame to the Brazilian goalkeeper Barbosa, but what about giving some credit to the Uruguayan goal-scorer Alcides Ghiggia? Ghiggia had also dribbled past Mustache to set up the tying goal.

35. National soccer federations can be just as corrupt and, in the case of Brazil banning Barbosa, even more illogical than FIFA.

36. Racism also played a part in the ostracism of Barbosa and Bigode. Both were accused of being intimidated by Uruguay—an accusation leveled against all the black Brazilian players.

It wasn't just Ghiggia. Uruguay were a great team. By defeating Brazil in 1950, Uruguay had won two of the first four World Cups, and they hadn't even entered the other two.[37] Brazil, on the other hand, were playing in a World Cup final for the very first time.

Two months before the final in 1950, Brazil and Uruguay played a series of three games in Rio; Uruguay won one of the three. In seventeen games against Uruguay since 1938, Brazil had lost five times. So perhaps Brazil were a slight favorite in that final, but their victory was far from a sure thing.

Asking the series of *why* questions uncovers reasons for Brazil's loss. And if the objective is to improve Brazil's chances in future World Cups, it's better to ask a few more questions than to rely on scapegoating the goalkeeper.

Asking the five whys helps us connect dots, test theories, and uncover perspectives we might otherwise miss.

So, while counterfactual analysis is a systematic way to envision alternative possible outcomes, asking the five whys is a systematic way to uncover possible root causes.

The five whys can also help us understand non-soccer Brazilian tragedies. The northeast region of Brazil is one of the most underdeveloped areas in the Western Hemisphere and is a place where some children routinely eat dirt. (I also ate dirt as a kid, but my parents tell me it was a phase, not a necessity.)

> *Why do children in northeastern Brazil habitually eat dirt?* The children lack iron in their diet, so they instinctively eat dirt to get the mineral salts needed to stave off anemia.
>
> *Why does their diet lack iron?* Because the foods that have iron cost too much.
>
> *Why do foods that have iron cost too much?* One reason is that many people in northeastern Brazil are economically poor. The few jobs rarely pay well. But economically poor people can afford a balanced diet elsewhere. The effects of poverty are amplified in northeastern Brazil by high prices for food, prices as high as in affluent nearby beach towns. Not much food is produced in northeastern Brazil, and imported food has higher transportation and storage costs.

37. In those days, teams from South America were less likely to travel when Europe hosted the World Cup, and vice versa. When Uruguay won the bid to host the first World Cup, none of the five losing bidders (Hungary, Italy, Netherlands, Spain, and Sweden) came to the tournament. When Italy hosted the tournament in 1938, they were snubbed by Uruguay (and by Colombia, Costa Rica, Dutch Guiana, El Salvador, Mexico, and the United States).

Why isn't much food produced in northeastern Brazil? The region does not have an inherently harsh climate, like Africa's Sahara desert or the southwestern United States. In fact, northeastern Brazil used to have some of the most fertile land in the world—most of the world's sugar was produced there through the mid-seventeenth century. The reason why not much is grown in northeastern Brazil is that the soils are worn out.[38]

Why are the soils worn out? A scarred landscape is the price people now are paying for mid-seventeenth-century sugar production that led to prosperity (for the rich people).[39] What used to be diverse forests with rich soil were burned into fields for sugarcane, which soaked up the soil's nutrients faster than they could be replenished. The once fertile soil is still not suitable for growing much, and the entire region suffers.

I'm instinctively good at figuring out what's wrong (and then complaining about it). The five whys force me into a more careful evaluation of what's happening in a system, which shows me unexpected and influential places to intervene for sustainability. I seldom find a magical "right" answer, but I always know more after asking more whys.

Some Brazilians were content to condemn Barbosa and Bigode. But by asking a few more whys, others realized that a better way to avoid future World Cup shame was to develop thousands of the best players in the world, like my friend Ze. In the same way, trying to fix children who eat dirt is missing the point; we need to return nutrients to the soil.

Brazil has made progress in both areas. In soccer they responded to their 1950 loss by making their team better. By the 1958 World Cup, Brazil's national team had even more talented players and was among the first to use a dietician, dentist, and psychologist. Brazil won that World Cup and now has won five in total, which is more than any other nation.

And people in the northeast region are now producing more diverse agriculture. They have learned how to grow a variety of crops, which

38. Not only is worn-out soil a poor growing medium, but it also sequesters less carbon than healthy soil.

39. The Dutch West India Company, the same one that bought/stole Manhattan from the Native Americans, took control of northeastern Brazil in 1630. They made most of the money from the new sugar plantations, and by 1654, when they were thrown out of the region, they had already established a competing industry in Barbados. The competition dropped prices. Slaves revolted. There was a gold boom in southern Brazil. These factors all contributed to the economic downturn in the region.

5

refreshes the soil and reduces the need to import food.[40] The infant mortality rate in the region is about half what it was as recently as 2000. Children in the northeast get a healthier start and, as it was for me, eating dirt is a choice instead of a necessity.

Asking the five whys can show us what is obstructing sustainability in the existing system. The five whys approach enriches our understanding of why children eat dirt and why Brazil lost in the 1950 World Cup final.

FOOTPRINTS OF THE WORLD CUP

Life-Cycle Assessment

To get ready to host the 2014 World Cup, workers in Brazil updated stadiums all over the country, including the "Estádio Garrincha," which

40. Mixed-crop agroecological farming like that in northeastern Brazil is spreading all over the world. In addition to adding to crop diversity, this type of farming can eliminate chemical pesticides and fertilizers while reducing water use with water-capture and rain-harvesting approaches. Crop yields decrease at first, but after a few years they increase as the soil comes back to life.

was named for our bent-legged angel. Garrincha's remodeled stadium got a new roof that filters air pollution[41] and collects rainwater for reuse. And it now boasts enough solar panels to power the operations of the entire stadium and send leftover clean energy into the power grid, where it is used by people in the surrounding city of Brasilia.

The next two World Cup hosts are following Brazil's lead. As Russia prepares for 2018, workers there are building six new stadiums designed to achieve green building certification. Qatar plans to showcase new solar-powered, net-zero-energy stadiums when they host the World Cup in 2022.

Sure, this is all nice progress, especially for those who get paid to design and build the stadiums, but the energy used in the stadiums is just a tiny fraction of the energy used to stage a World Cup. Researchers did a feasibility study to see what it would have taken to make the 2010 World Cup carbon neutral, limiting the carbon dioxide released into the atmosphere and then removing at least as much as was released. They found that energy to build and occupy stadiums was only 1 percent of the event's total energy consumption and far less than the energy used for international travel (67 percent), travel within the country (19 percent), and housing for visitors (12 percent).[42]

A zero-energy stadium does not make a zero-energy World Cup. If FIFA's only concern were the energy required to stage the event, they could choose a centrally located host country, which would minimize travel for fans. In fact, FIFA could encourage us to watch on television or the Internet—then fewer fans would travel in the first place.[43]

The greenwashed World Cup is not unique; most products and services claiming to be "green" mislead consumers about their environmental benefits.[44] Products boasting that they are "all natural" greenwash

41. The roof membrane technology uses ultraviolet light and a chemical reduction process to decompose nitrogen oxide and sulfur oxide, which are pollutants from exhaust fumes that produce smog and acid rain.

42. The "Feasibility Study for a Carbon Neutral 2010 FIFA World Cup in South Africa" was done by South Africa's Department of Environmental Affairs and Tourism in collaboration with the Norwegian Embassy.

43. Of course, FIFA would never tell fans not to travel to the World Cup and spend their money. But, if FIFA did want people to stay home, they could schedule the tournament somewhere fans are less excited about traveling to. Perhaps the criminal investigation of FIFA will reveal that this is the real reason Qatar was chosen to host in 2022.

44. Of the more than five thousand products in the "Sins of Greenwashing" study, over 95 percent mislead consumers about environmental benefits.

with vagueness (arsenic is naturally occurring). The hairspray advertised as "chlorofluorocarbon-free" greenwashes with irrelevance (chlorofluorocarbons are illegal to begin with). The Hybrid Cadillac Escalade sport utility vehicle greenwashes by advertising the lesser of two evils (it may be a hybrid, but it is still far less fuel efficient than other vehicles).

Greenwashing is a problem because when we incorrectly think we are doing better, we get lulled into complacency. A 2014 Ford Explorer has about the same fuel efficiency as a 1908 Model T (seventeen miles per gallon). "Green" buildings, on average, use more energy per square foot than buildings built ten, fifty, or a hundred years ago.

Don't get me wrong. Remodeling Garrincha's stadium to be zero-energy is better than remodeling it to use a lot. But we need a way to judge potential steps on our quest for sustainability.

To evaluate and prioritize different actions, we can apply life-cycle assessment (LCA), in which we consider the benefit provided by a product or service and the resulting sustainability impacts over a period of time.

Setting up an LCA is like defining our systems: we specify our scope, impacts, and functional units.

The scope of an LCA typically reflects a system from raw material extraction through processing, manufacture, distribution, use, repair and maintenance, and disposal or recycling or reuse. For example, recall the LCA for the (impossible) sustainable hamburger, which was scoped to include raising cows; butchering and processing the cows; transporting the hamburgers to consumers; and storing, cooking, and disposing of the hamburgers.

In the feasibility study for a carbon-neutral 2010 World Cup in South Africa, the scope was the event itself and the build-up to it. The study was limited to benefits and impacts for those attending in person. Assessment of the stadiums considered details such as the energy used to manufacture steel, transport it to the stadium, and then hoist it into place. Of course, not all details can be captured; the assessment did not include the carbon emissions associated with all the McDonald's hamburgers the construction workers ever ate.[45]

By defining the scope in this way, the feasibility study avoided an overly narrow focus on stadiums as the best way to make the 2010 World Cup carbon neutral. Such a mistake might have led to greenwashing through hidden trade-offs, claims that something is more sus-

45. A rule of thumb is that impacts making up less than 1 percent of the total can be excluded. That said, you don't know for sure until you measure it.

tainable based on a narrow set of attributes (World Cup stadiums) without attention to more important ones (World Cup travel).

In addition to selecting the timeframe and level of detail for our LCA, we also choose which impacts to consider. We might consider financial impacts to check whether money spent on the event would do more good if it were spent on other projects. We might evaluate social impacts, counting the number of people who are displaced from their homes by new stadiums and probing the treatment of migrant workers building the stadiums. Carbon dioxide emissions were the impact of interest in the feasibility study of the 2010 World Cup.

We can define scopes and impacts, but the LCA is useless if we pick the wrong functional unit, which is basically the product or service being provided. Defining a functional unit lets us compare between alternative scenarios.

The functional unit in our hamburger example is the amount of nutrition provided. An easy way to halve the embodied energy of the hamburger would be to make it half the size, but that would only provide half the nutrition. Therefore, our functional unit would remind us that the half-sized hamburger is no better (in its ratio of nutrition to embodied energy)[46] than the regular hamburger.

For the World Cup, a reasonable functional unit might be the amount of happiness fans take from the event. Thus, a World Cup that requires a little bit more energy but brings happiness to a lot more fans may actually be preferable, based on its energy-to-happiness ratio.

With the system boundaries, impacts, and functional unit defined, the next step of the LCA is to take an inventory of flows into and out of the system. Flows might be water, raw materials, happiness, embodied energy, CO_2 emissions, and so on. In more and more cases, we can get data for these flows from public databases. Other times, we may need to gather more specific information through field measurements, surveys, experiments, or interviews.

Once the inventories are taken, impacts can be categorized and combined into common units—whatever helps us interpret the results. Our interpretation is more accurate the more we know about how the LCA was conducted. It helps to know which data are contributing most to impacts—and the sensitivity of those data. For example, much of

46. Most of us would survive if we ate smaller hamburgers. And if we all ate half the amount of hamburgers and got the rest of our nutrition from a lower-carbon source, that would be a massive sustainability improvement.

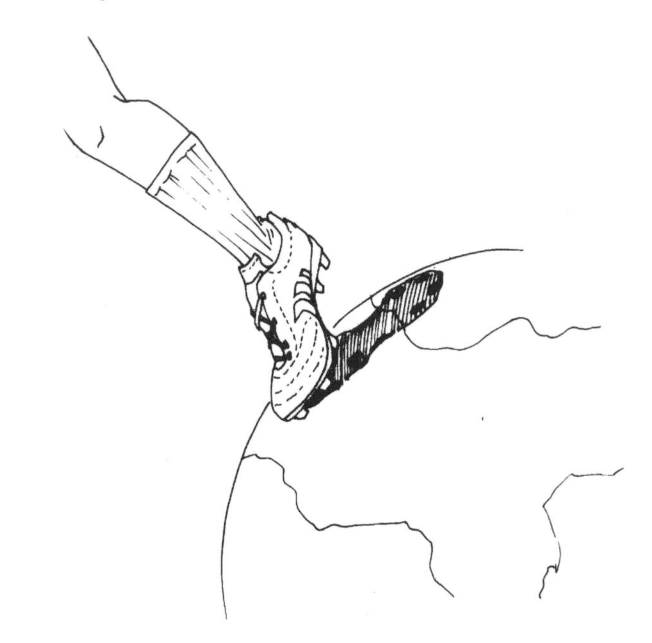

the 2010 World Cup's CO_2 impact comes from air travel. So, if air travel turned out to be much better or worse than originally estimated, we could then evaluate whether that new insight meaningfully changed the overall findings in the feasibility study for a carbon-neutral World Cup.

The LCA approach underpins any self-respecting "footprint" measurement, which is a catchall term for various types of human demands on Earth's ecosystems. What's nice about footprints is that they put all the technical detail about the LCA in the background and let us focus on what the footprint means. For our personal footprint, we are the functional unit, the scope is our personal use, and the impact can be measured in terms of energy, CO_2 emissions, environmental degradation, and so on.

I've spent hours using Internet calculators to estimate various footprints, usually when I'm contemplating some new behavior such as substituting nuts for meat in my diet, driving less, or buying only used clothes. My results depend on the underlying method in the footprint calculator and the input I provide.

Measuring my footprint can be discouraging because, no matter which calculator I use, the results show that I need to reduce my footprint by at least 75 percent to make it so that everyone could live like me without exceeding our planet's carrying capacity.[47] This is in spite of the fact that I only traveled to watch the World Cup in 1994, when the games were played so close to home that we drove, with five people in one non-Escalade vehicle.

But I'm encouraged by footprints when I imagine everyone measuring theirs. I imagine people striving to be remembered for reducing footprints instead of chasing temporary immortality by building monuments to themselves. I imagine estate planners becoming as knowledgeable about environmental inheritances as they are about economic ones. I imagine more people like my parents, who, after putting three kids through youth soccer and college, finally had some money for themselves, and spent it not on a second home, but on an array of solar panels that will provide enough clean energy to more than offset their already small footprints.

Life-cycle assessment can shape planning for more sustainable World Cups. The same LCA approach also lets us objectively reflect on our own behavior, which is why I promise that calculating your footprint will change your life.[48]

Life-cycle assessment is an objective way to evaluate and compare sustainability among different scenarios. These assessments show us that trying to green a World Cup via the stadiums is futile and that we need to discover a few more habitable planets for everyone to live like me (and probably you).

47. I keep an annually updated record of my carbon footprint on my website (http://essoresearch.org/offloading-my-carbon-footprint/).

48. My current favorite footprint calculator is the CoolClimate Network's calculator (http://coolclimate.berkeley.edu/calculator), which strikes a nice balance between detail and usability.

Review: Evaluating

Path dependence is when small initial differences in a system are a disproportionate cause of differences in outcomes. Or, in other words, history shapes the present.

Inertia is a resistance to change in the state of motion—no different from the inertia concept I had forgotten from my first physics course. Inertia is why systems at rest tend to stay at rest and why systems in motion tend to stay in that same state of motion. When we know the state of motion, we can use it to our advantage instead of hopelessly fighting it.

Carrying capacity is the population that an environment can sustain indefinitely, given the available resources and the rate at which they are being consumed.

Counterfactual analysis is used to study alternative outcomes. It's basically the same after-the-fact second guessing that we do all the time, except it's more methodologically sound.

Asking the five whys helps us connect dots, test theories, and uncover perspectives we might otherwise miss.

To evaluate and prioritize different actions, we can apply life-cycle assessment (LCA), in which we consider the benefit provided by a product or service and the resulting sustainability impacts over a period of time.

6

Creating

Once we've defined and evaluated our systems, we're ready to create systems that are more sustainable.

TREES IN MAPUTO AND THE SACRED MONSTER IN AMSTERDAM

Overview: Creating Sustainable Systems

The stealth goal-scorer, Eusébio, is the Black Panther. The powerful midfielder, Coluna, is the Sacred Monster.

Maputo, the Mozambique port city that gave us Eusébio and Coluna, has a nickname too. It is called the "City of Acacias" for the trees that line its streets. The acacias amplify Maputo's beauty, in particular when their leaves turn red. And for those poor souls unmoved by beauty, well, the trees filter the air and provide shade too.

Acacia trees even give us ideas for more reliable drinking water in Maputo. Sometimes Maputo simply gets too little water, as was the case during droughts in 1999. Other times, the problem is too much undrinkable water, when there are pollution-spreading floods like those that occurred in 2000 or when rising seas cause salination. Acacias persist through these changing conditions thanks to long tap roots that reach deep to dependable groundwater and also filter contaminants.[1]

1. The xylem tissue in trees has pores that allow water to pass through but that trap most bacteria. In fact, pouring water through a freshly peeled pine tree limb has been shown to remove most bacteria; see Boutilier, M.S.H., Lee, J., Chambers, V., Venkatesh, V., and Karnik, R. (2014), "Water filtration using plant xylem," *PLoS ONE*, 9(2), e89934.

Of course, a single acacia tree on a Maputo street corner does not think to itself "I better grow a tap root deep into the ground;" the trait came from round after round of reproduction in which the trees best suited for the conditions survived to pass along their genes.

We tap into millennia of evolutionary wisdom when we practice *biomimicry;* we observe how Nature is already doing whatever it is we are trying to do, and then apply what we find. Using biomimicry often leads to a related approach to creating sustainable systems: *closing loops,* which means eliminating waste and ending dependence on inputs from outside our system. Dead acacia leaves, for example, aren't a waste. The leaves don't need to be hauled to a landfill or even to a recycling center. When leaves fall off the acacia, they decompose and provide nutrients to the soil and to the growing tree. As we'll see, biomimicry and closing loops can help whether our goal is sustainable water in Maputo, more efficient wind turbines, or better scoring chances on corner kicks.

Whether acacias or people, whatever hopes to live in Maputo must adapt to the floods, droughts, and other changes in the climate.[2] In Maputo, as in most cities, the challenge is compounded because the parts of the city where these hazards hit hardest are home to the most vulnerable people. A mother in Maputo's slums cannot move her family to a hotel and wait out the flooding. Her husband is not choosing which brand of bottled water to buy. Their decision is either withhold polluted water from their children and hope they don't die from dehydration, or give their children polluted water and hope they don't die from cholera.

Whether the family survives depends on their *adaptability,* their opportunities to adjust to disturbances by modifying behavior. When the family has access to several wells, they can survive the contamination of one. When the family is connected to a reservoir-supplied water system, they are buffered from droughts and salinated groundwater.

A process that leads to adaptability, and to sustainability in general, is *polycentrism.* In Amsterdam in 1962,[3] the Sacred Monster and his Benfica team were back in their second consecutive European Cup final.

2. The drought/flood pattern is widespread, including in the northeastern Brazilian region we discussed previously. Unpredictable climate makes subsistence farming in the region more risky, and displaced farmers end up crowded into the favelas of Rio and São Paulo.

3. A fifteen-year-old Johan Cruyff was a ballboy (not smoking) for that European Cup game in Holland. How's that for interdependence!

The Sacred Monster had just scored to tie the game when Benfica were awarded a penalty kick (bear with me; this ties back to polycentrism). Through his leadership and successful penalty record, the Sacred Monster had earned penalty-taking duties for Benfica. Yet, in the biggest game of the season, the newcomer Eusébio asked to take the kick instead.

Coluna let him.[4]

And Eusébio scored.

As captain and designated penalty-taker, the Sacred Monster could have refused the request and scored the penalty himself. But the captain instead practiced a sort of polycentrism, balancing management and autonomy in order to manage a shared resource. Time and again, people linked by common interests collaborate to sustainably manage common resources like water (or penalty kicks).

Collaboration can't happen without *transparency:* access to the right information, in the right format, at the right times. Agencies in Maputo encourage polycentric water management by sharing and seeking input on design alternatives. As a result, designers are not just the professionals; designers are anyone trying to change what exists into what they want.

A transparent and polycentric process leads to more sustainable systems, and so does *correcting systematic inequality,* or removing imbalances that are perpetuated by the system. Some inequality is inevitable and useful; we don't want every game to end in a tie. Still, when the same team always wins, eventually they have no one to play with. When there is systematic inequality, everyone suffers. Maputo's floods devastate the day-to-day lives of people in the slums, but no part of the city is sheltered from the economic impact.[5]

After converting the go-ahead penalty in that 1962 final, Eusébio scored again to clinch Benfica's second straight European Cup. Eusébio's stunning goal was possible because of a simple pass from the Sacred Monster.

4. Legend has it that when the penalty in that 1962 European Cup final was called, an opponent called Eusébio a *maricón.* So, when Coluna agreed to let Eusébio take the penalty, the captain also told his protégé (who didn't understand the insult) to make the penalty and then call the opponent a *cabron.* (If you don't already know, you can find the meanings of those insults on urbandictionary.com.)

5. The floods in 2000 led to an estimated $1 billion dollars of lost economic activity.

The pass exuded Coluna's *elegance,* defined as the surprising simplicity that is beyond complexity. An elegant design for clean water in Maputo might mean scrapping plans for an underground sewer system with a central treatment facility, in favor of raised pit latrines. This low-tech approach would better serve the vast majority of people in Maputo without sewer connections. An even more elegant approach might mean investing in hygiene education instead of physical infrastructure.

Like me, you are probably learning about sustainability because you want to somehow make things better, whether it's providing a better life for kids in your community, avoiding the worst effects of global climate change, or helping your team win the league. Creating is how we pursue our sustainability quests, and the ideas in this section can help ensure that we are heading in the right direction.

FALSE-BROODING RUNS AND WIND TURBINES

Biomimicry

After growing to 1.9 meters tall, my "little" brother usually towered over his soccer opponents. Tall players have an advantage on corner kicks when the player taking the kick passes the ball in the air in front of the goal. Attackers try to head the ball past the goalkeeper, and defenders try to head it clear.

So, more specifically, players with heading ability have an advantage on these corner kicks. Heading ability tends to correlate positively with height—although not for my brother.

As a corner kick is taken, attacking players disperse to positions from which they could score if the ball comes to them. "If the ball comes to them" is the key, because the corner-kick-taker aims for a specific spot. Runs away from this intended target are known as "dummy runs" (it's insensitive to dummies to compare them to my brother, so I'll propose another name for these runs soon).

Despite his height, my brother was not the target on most corner kicks. He would typically sprint to the goalpost closest to the side the ball was on, pretending that he thought he would get the ball, when in reality he knew that the passer was aiming elsewhere. If this dummy run worked, it would lure my brother's defender away from the intended target. And because my brother's height made him look like a corner-kick-scoring threat, opponents often wasted one of their better defenders guarding him.

My brother's dummy runs meant that he had fewer chances to score, but they were better for his team.

I propose we rename the dummy run the "false-brooding run" to more accurately describe my brother's role. False brooding is a trick that birds like plovers use to protect their young from snakes.[6] A false-brooding plover pretends to sit on a nest away from its actual nest and then, when a snake approaches, flies away at the last second, leaving the snake far from the intended meal of plover eggs. My brother runs where the ball won't come to distract the defender;[7] the plover broods where there are no eggs to distract the snake.

I'm not sure who first used false-brooding runs in soccer, and they probably were not inspired by plovers. But if they were, it would have been an example of biomimicry.

Biomimicry means asking ourselves what in the natural world already does what we are trying to do, and then learning from that example to make our approach more sustainable.

Nature will speak—we need to be listening. Dragonflies were airborne more than three hundred million years before the Wright brothers. The trees in Maputo filter polluted water. And foraging ants don't get stuck in traffic.

Among the many contributions of biomimicry[8] is lowering the costs of wind energy. A renewable resource, wind energy is far better for the environment than fossil fuels, in terms of climate-changing emissions in particular.[9] Still, wind energy remains only a fraction of our energy

6. To learn more about the intricacies of false brooding, and perhaps even pick up some corner kick strategies, start with Walters, J.R. (1990), "Anti-predatory behavior of lapwings: field evidence of discriminative abilities," *Wilson Bulletin*, 102(1), 49–70.

7. Like my brother, I'm tall for a soccer player, but I was rarely targeted for headers on corner kicks. Instead, I would fake like I was going to run for a header but really set myself up for a shot with my feet. In other words, my run would be like a plover making a distraction—and then circling back in search of something to eat for itself.

8. Biomimicry is broadly applied in the field of industrial ecology, where healthy ecosystems serve as a model for more sustainable human socioeconomic systems.

9. The 2014 Intergovernmental Panel on Climate Change calculated the life-cycle CO_2 equivalent of various electricity-supply technologies. Per unit of energy produced, wind produces about forty times less CO_2 than natural gas and seventy times less than coal. The relatively small amounts of CO_2 emissions associated with wind come from the construction and transportation of the turbines themselves (search for "IPCC Working Group III—Mitigation of Climate Change, Annex II I: Technology-specific cost and performance parameters").

supply,[10] mostly because wind has higher economic costs than fossil fuels.

One way to improve the economic case for wind energy is by increasing the efficiency of the wind turbines. If a different turbine design collects more wind energy for the same production cost, then the price of wind power comes down, which makes wind a more attractive choice for electricity companies and their customers.

Applying biomimicry to improve turbine efficiency, we would seek an example from Nature that is accomplishing a similar purpose.

We would ask something like "What in Nature efficiently moves through a fluid?"

And we'd get plenty of answers.[11]

One surprising example is the humpback whale's flipper, which has big bumps across its leading edge. It turns out that the bumpy flipper has better fluid dynamics than our smooth turbine edges. Wind tunnel tests confirmed that turbines with bumps have less drag, and companies are making bumpy turbines.[12]

We find additional opportunities to better harness wind energy when we take a broader view. Most wind farms consist of turbines that spin parallel to the pole; these horizontal-axis turbines collect more energy per turbine. On the other hand, vertical-axis turbines, which spin around the pole, collect less energy per turbine but can be spaced closer together, which means that more fit on the farm.

To enhance spacing of vertical-axis turbines, researchers learned from configurations of schooling fish, which use less energy moving through water in a group than they would if they were swimming alone. What works for fish could work for wind farms. Preliminary results suggest that farms with vertical-axis turbines spaced like fish schools could outperform wind farms with conventional spacing.[13]

10. In 2015, just 4 percent of electricity in the United States came from wind; 67 percent came from fossil fuels (the Energy Information Administration has these statistics and many more; www.eia.gov). Worldwide, the fraction provided by wind is even smaller (search for the World Bank's "World Development Indicators").

11. The website AskNature (www.asknature.org) is a great resource for biomimicry inspiration. For instance, I did a quick search and found that Adidas's Jabulani designers could have learned from barracuda mucus and owl feathers to avoid the turbulence that made their ball unpredictable.

12. The clever company name "whalepower" is already taken (www.whalepower.com).

13. Field experiments of the wind arrays arranged like schooling fish have been less promising than theoretical results, but still an improvement over single arrays or

Just as false-brooding plovers explain my brother's running on corner kicks, Nature shows us ways to better harness wind energy. And plenty of mimicking opportunities remain. The humpback whale's fin and the school of fish perform multiple functions and are built by self-assembly with entirely recyclable materials, just like everything else in Nature. Perhaps one day we will be able to say the same for wind turbines.

It makes sense to learn from Nature, which has been around for a while. My brother acted like a plover on corner kicks, and biomimicry of whale fins and schooling fish has led to more efficient wind power.

SOCCER-SHIRT QUILTS AND THE RECYCLING DISTRACTION

Closing Loops

My mom was a "soccer mom" long before it became a demographic.[14] She's watched[15] at least fifteen hundred games in which my sister, my

unplanned spacing. See Dabiri, J. O. (2011), "Potential order-of-magnitude enhancement of wind farm power density via counter-rotating vertical-axis wind turbine arrays," *Journal of Renewable and Sustainable Energy*, 3(4), 043104.

14. During an election for Denver City Council in 1995, Susan Casey ran with the slogan "A Soccer Mom for City Council." Then, in campaigns for the 1996 U.S. presidential election, soccer moms were sought-after swing voters. The term has been overused ever since.

15. My mom always watches from the offensive side of the field because that helps her kids score and their teams win. When the teams switch ends of the field at halftime, so does my mom.

brother, or I played. That's five full months of her life! And it's probably closer to five years if we add in all the driving, team meals, stays in shabby hotels with shabbier teenagers, and so on.

But it's not all a selfless sacrifice, because as the soccer mom invests time, she does get some personal rewards. My mom, for example, can now balance a ball on her foot. And she knows that trying to lure the opponent offsides without putting defensive pressure on the ball is foolish.[16]

The graduation ceremony for soccer moms is to create a quilt from their kids' old soccer shirts. It's a rite of passage after which the mom might even recall that she has her own interests. But first, as with the rest of the soccer-mom experience, making the shirt quilt means the mom does the work and the kid gets the benefit.

The quilt my mom made for me has pieces of my soccer shirts from when I was an immature ten-year-old amateur and from when I was an immature twenty-two-year-old professional, and everything in between.[17] Instead of being trashed or stored unused in a closet, the shirts get a second life as part of a quilt that keeps us warm and reminds us of our soccer stories and of the unpayable debt we owe our soccer moms. The shirt quilt closes a loop for soccer moms and for the life cycle of the shirts.

Closing loops means eliminating waste and ending reliance on inputs from outside the system. (It's also a type of biomimicry.)

Closing loops is why I cringe each time well-intentioned people jump straight to recycling as their go-to initiative for sustainability. Recycling is just one of countless ways to close material loops, and not typically the best. For instance, in North America, only about 1 percent of material flows end up in a product that is still being used six months after it is sold.[18] In other words, recycling my plastic chocolate wrapper recovers only a tiny fraction of the materials that have been disturbed for me to eat chocolate.

16. Fortunately for me and my college teammates (twice in one season), one of our opponents did not appreciate that an offsides trap is easy to beat when there is no pressure on the ball.

17. As soccer moms know, choosing which shirts to include is only slightly easier because many are dirtied far beyond launderability.

18. This estimate of wasteful material flows is in the book *Natural Capitalism: Creating the Next Industrial Revolution* (it's free online: www.natcap.org/sitepages/pid20.php).

What's more, recycling is a big loop. And the bigger the loop, the harder it is to close it because more inputs are needed and more outputs are produced. When I put my plastic chocolate wrapper in the recycling bin, I don't truly close a loop, even for that tiny fraction of my chocolate-eating experience. The recycling truck uses gas to come empty the bin and take the contents to a sorting facility. Then another truck takes my wrapper and the similar types of plastic to a factory where everything is smashed and shredded and, maybe after being transported again, reconstituted to make something else. Each step uses energy and water and releases climate-changing pollutants.

Of course, recycling is better than throwing my chocolate wrapper in the trash, in which case it would still be occupying landfill space during the 2446[19] World Cup. And recycling beats littering the wrapper on the beach and adding it to the plastic-strewn vortex accumulating in the North Pacific Ocean. So, I'm not going to stop recycling (or eating chocolate), but I know I'm not saving the world when I do so. My negative impact is a tiny bit smaller, but I am far from closing the loop.

Closing loops is why the advice "Reduce, Reuse, Recycle" is ordered the way it is.

One way to reduce is through dematerialization, by replacing products with services. For example, I don't really care about the physical product that is the chocolate packaging. I just care about the service: the chocolate reaching me undisturbed. Providing this same service while eliminating or reducing the physical packaging has the most potential; it could eliminate the entire recycling-process loop.

After reducing the packaging, reusing it is the next best option. Reusing eliminates the transportation and remanufacturing needed for the recycling process.

And finally, in the quest for sustainability, recycling is a last resort—not the first choice.

19. As of this writing, FIFA has not announced who will host the World Cup in 2446.

Closing loops creates more sustainable systems by reducing both waste and reliance on inputs. The soccer-jersey quilt closes a loop for soccer moms, and recycling closes a material loop—but not as well as reducing or reusing.

THE GOALKEEPER PICK TRICK AND IRISH LUMPERS

Adaptability

My mentor[20] in professional soccer, Doug Petras, taught me the goalkeeper pick trick when I was playing[21] for the Harrisburg Heat indoor team. Unfortunately, I never got to use the trick myself, which is one of my great soccer regrets. So, yes, this section is about adaptability for sustainable systems, but I also hope to promote the goalkeeper pick trick.

One day in practice, our coach had us pick teams for a scrimmage among ourselves. As usual, Doug was picked first. He was our starting

20. My most memorable mentoring advice from Doug was when he learned that I had graduated with an engineering degree and asked me, "Why are you playing soccer?"

21. The Heat's general manager wisely wrote my contract to pay me only when I got into games, which happened just five times, and when I scored, which happened once (as astute readers will recall from when I compared myself to the legend Eusébio.)

goalkeeper, and so much better than his backup that Doug's team literally never lost in any of about twenty scrimmages we played that season.

For this particular scrimmage, Doug secured the outcome even before it began. After the other team used their first pick on one of the best field players, Doug persuaded the captain who had just picked him to select the backup goalkeeper with the very next pick. Everyone thought Doug had wasted a pick on a redundant goalkeeper.

But then we realized that the other team was stuck without a true goalkeeper. There was a massive drop-off in performance from Doug to his backup. But it was nothing compared to the drop-off from Doug's backup to any of the non-goalkeepers on our team. Not having a true goalkeeper is a handicap in any type of soccer, and it's debilitating in the version of indoor soccer we were playing, which is basically ice hockey with much bigger goals.[22]

Doug's goalkeeper pick exploited field players' lack of adaptability to the goalkeeper role. A similar trick wouldn't work with any other position; if Doug's team had picked all the defenders, for example, the other team could have just moved some adaptable midfielders back to cover those positions. In fact, one reason our manager let us pick our own teams was to see whether unexpected new roles would emerge for field players.[23]

Adaptability is being able to adjust to disturbances by modifying behavior—without any outside intervention. Diverse systems are more adaptable.

The lumper potato was wiped out by a disease in mid-nineteenth-century Ireland. Lumpers in other European countries were wiped out by the same disease, but the effects weren't as severe because the people in those countries could adapt—they had other varieties of potatoes, and different types of food. The Irish had few other options to meet nutritional needs, so the lumper disease led to famine and exodus.[24]

Dependence on single crops, whether directly for food or as a dominant trade export, may suffice in the short term, but dependence ultimately restricts autonomy and adaptability. Or, as the Cuban poet/

22. We played five on five with unlimited substitutions, rapid-fire shots, and scores that typically reached double digits—especially when one of the goalkeepers was a converted field player.

23. After Doug's pick trick, the goalkeepers always picked the teams to maintain at least some semblance of competition.

24. Many who survived the famine did so by leaving Ireland. Or, in systems terms, the migrants expanded boundaries beyond national borders to include places where there was adaptability in the food supply.

philosopher José Martí put it, "A people that entrusts its subsistence to one product alone commits suicide."

Single-product prosperity can be interrupted by disease, competition, or environmental degradation.[25] The Irish potato experience has been mirrored by cacao in Venezuela, cotton in northeast Brazil, tannins in northern Argentina and Paraguay, agave in Mexico's Yucatan region, and coffee in too many places to list.[26] When the prosperity ends, the local growers dependent on the monoculture are stuck with scarred land and without income. The colonizers and companies, both at least as responsible as the locals for the monoculture, can move on to exploit the next boom.

Redundancies are a Band-Aid for dependency on a single product. We can delay the need to adapt by stockpiling resources, whether goalkeepers, food, money, or oil. But such an approach is wasteful because redundancies, by definition, require extra resources.

Diversity, by contrast, is a less wasteful approach. Diversity builds adaptability by adding to the number and variety of relationships and approaches. If there had been more variety in Ireland's food system, or just more variety in their potato varieties, the famine would have been less devastating. If we could meet our energy needs with a balanced mix of wind, solar, nuclear, fossil fuels, and biofuels, we would be less susceptible to disturbances in any one area.

25. A side effect of monocultures is destruction of the biodiversity that works behind the scenes to replenish food and freshwater, buffer us from natural hazards, and build our resistance to pests and pathogens.

26. For more on this history of booms and busts in Latin America, I recommend a non-soccer book by Eduardo Galeano: *Open Veins of Latin America: Five Centuries of the Pillage of a Continent.*

And if any of the field players on our team could have played goalkeeper at a professional level, Doug's goalkeeper pick trick would have backfired.

Adaptable systems perform multiple functions, which lets them respond to unpredictable but inevitable shocks. In agriculture, monocultures such as Irish potatoes are not sustainable because they cannot adapt. In soccer, the goalkeeper pick trick exploits limited adaptability.

"LOS GALÁCTICOS" AND NEW ENGLAND LOBSTERMEN

Polycentrism

Around the time when I was playing soccer for $2,000 a month, the Spanish club Real Madrid were assembling a team of world superstars that came to be known as "Los Galácticos."[27] In 2000, Madrid paid $60 million for the Portuguese winger Luís Figo. The following year, Madrid shelled out $70 million for the French midfielder (and our favorite threshold-crosser) Zinedine Zidane. In 2002, it was $45 million for the Brazilian forward Ronaldo. Then, in 2003, Madrid paid $38 million for the English midfielder (and husband of Posh Spice) David Beckham.[28]

Vicente del Bosque was tasked with managing Los Galácticos, and he faced an extreme version of a challenge faced by every professional soccer manager. The manager gets paid, first and foremost, to do whatever gives his team the best chance to win. But a player's value depends on far more than the team's record. It matters how much they play, how many goals they score, whether they stand out, whether they get to take penalty kicks, and so on.

Aligning team and individual purposes is even more challenging because professional managers often make less money and have less job security than the players they are supposed to manage. If the team is losing or if players (especially the most expensive ones) are unhappy, the manager is usually the first to go.

Del Bosque knew that Los Galácticos would never reach their potential if he made all the decisions while the players did all the work.

27. "Los Galácticos" translates as "The Galactics"—glad I could help.

28. Ronaldo, Zidane, and Figo won World Player of the Year seven times between them. David Beckham never won World Player of the Year, but he did help Madrid sell plenty of jerseys in Asia and North America.

So, even though Del Bosque was on the hook for the team's performance, he sought players' advice on training methods and tactical approaches.

With input from the superstars, Del Bosque figured out a way to get both Figo and Beckham on the field at the same time, even though the two had played the same position on their pre-Madrid clubs. Individual and team needs were met as Del Bosque provided top-down integration and the stars provided bottom-up practical knowledge and motivation. The balance of individual autonomy and team order seeped into Madrid's on-field playing style, in which structured tactics were punctuated by spontaneous bursts from Los Galácticos.[29]

In Del Bosque's four seasons as manager, Real Madrid won two Spanish league titles and twice won the European Champions League.[30] It was the club's most consistent success since the 1960s (when they were battling the Sacred Monster's Benfica teams).

But there was a problem in Madrid.

The harmonious balance of autonomy and control between Del Bosque and his players did not extend to Real Madrid's then president, Florentino Pérez, who meddled in decisions that Del Bosque knew were best left to the manager. Del Bosque told Pérez so, which created a divide between the two, and despite overwhelming success on the field, Del Bosque was fired.

After Del Bosque, Pérez tried four different managers in four seasons. Pérez dictated tactics and pressured the managers to play the superstars, regardless of their fitness or form. In the four seasons after Del Bosque, with most of Los Galácticos still on the team, Madrid never won the Spanish League or a major tournament.

Pérez traded the practical knowledge of the manager and players for the detached agenda of a supposedly high-level decision maker, and Madrid paid the price as the superstars' motives eclipsed the team's purpose.[31]

29. Marcelo Bielsa, an Argentinean manager and soccer savant, explains the tension between autonomy and order in organized soccer: "Totally mechanized teams are useless, because they get lost when they lose their script. But I also don't like ones that only rely on the inspiration of their soloists, because when God doesn't turn them on, they are left totally at the mercy of their opponents."

30. In that 2002 European Champions League final, Zidane scored a stunning left-footed goal to help Madrid defeat Germany's Bayer Leverkusen. My Pittsburgh Riverhounds team had played Leverkusen on their preseason tour of the United States that year (we lost 6–4). That's the closest soccer connection I have with Zidane.

31. As a Barcelona supporter, Pérez's intervention was all right with me.

Polycentrism is a balance of management and autonomy at multiple levels to sustainably manage a shared resource.

Like the manager Del Bosque, political scientist Elinor Ostrom can teach us how to balance control and autonomy with polycentrism. Ostrom studied these ideas in resource stocks such as fisheries and woodlands.[32] These "common pool" resources present a dilemma because they can be so dispersed that users have no incentive not to use them at unsustainable rates until they are completely gone. If we don't eat the fish or cut down the trees, someone else will, the thinking goes.

The theory is that free access and unrestricted demand for a finite resource stock is a recipe for exploitation. Users get the benefits of the resource, while the costs are shared by everyone, regardless of whether they use the resource. So more and more people decide to use the resource, until it's eventually gone.

Ostrom found that the theory doesn't always match reality. Humans *can* effectively manage common pool resources. For proof, consider that my brother and his wife[33] served fresh New England lobsters at their wedding.[34]

Lobsters were once such a plentiful common pool resource that my grandmother, who is from New Hampshire, remembers lobsters as a low-priced staple food. She claims to have eaten so many that she got sick of them, although her appreciation had evidently returned by dinnertime at my brother's wedding.

As my grandmother grew up, the lobster population fell as demand increased.[35] As the population fell, lobsters commanded higher prices, which made lobstermen more eager to catch them. And they did, with more boats and new lobstering techniques.

32. *Bosque* is Spanish for "woodland"—which is ironic, or perhaps even an unexpected interdependency.

33. Yes, this is the same loving relationship that began with my brother dominating co-ed pickup soccer.

34. Effective management of the commons is only one of the reasons lobster fisheries are now better off in New England. Another likely reason is that competing species like cod have been overfished. Plus, the lobster fishery remains unstable as warmer oceans and acidification from climate change threaten lobsters in southern New England. And, as you know by now, the goal should be ecosystem-wide sustainable yields, and not just optimization of a single species. For more, see Howell, P. (2012), "The status of the southern New England lobster stock," *Journal of Shellfish Research*, 31(2), 573–579.

35. Demand increased in part because there were more people and in part because lobsters tasted better after lobstermen began transporting them alive to markets.

4

Lobstering communities rely on a consistent annual harvest, which was in jeopardy as the population flirted with endangered species lists. Even if some fleets limited their catch, lobster numbers would continue to plunge. In this scenario, the theory said that individual actors will try to get it while they can.

However, in one of her many studies of how people manage common pool resources, Ostrom learned that the New England lobstermen created rules and enforcement to ensure a sustainable harvest of lobsters. Just as Madrid's players balanced team and individual needs when Del Bosque gave them autonomy, the lobstermen sustained their fishery with minimal outside influence.

The lobstermen made simple rules. The first time someone caught extra or out-of-season lobsters, other lobstermen would wrap a bow tie around the violating lobster trap. On the second offense, the lobstermen visited the offender's home to discuss the problem. If it happened again, the lobstermen smashed the trap. If that didn't work, the lobstermen planned to sink the offender's boat—though no sinkings were reported.[36]

36. In Higuita's Colombia, poachers who disobey a fishermen-imposed ban on using nets during breeding season are not so lucky. Offenders there are wrapped in their nets and thrown into the river, according to Alan Weisman's book *Gaviotas: A Village to Reinvent the World.*

Ostrom found that common pool resources get managed more sustainably when there is both community control and larger-scale governance. New England lobstermen acted in part to avoid the restrictions that would have kicked in if their lobsters were added to national endangered species lists. In Madrid, the superstars knew that Del Bosque would give them leeway, but that they could also find themselves on the bench if they were not advancing the team's best interests.

After leaving Madrid, Del Bosque eventually was asked to manage the Spanish national team and, as he had at Madrid, he instilled a balance of structure and player autonomy. Del Bosque took over in 2008, and Spain won their next thirteen games, which remains a record for the most consecutive wins by a new national team manager. Spain even went on to win all ten of their qualifying games for the 2010 World Cup in South Africa.

Del Bosque practiced polycentric management across Spain, and Ostrom found it all over the world, from forests in Japan to irrigation in the Philippines. Polycentric approaches are now used for everything from water management in Maputo to urban development in Europe. Elinor Ostrom won the Nobel Prize in 2009 for her work in this field.

Just seven months after Ostrom accepted her prize, the Del Bosque–managed Spanish national team played in South Africa. There, Spain used the polycentric style to win the 2010 World Cup and prove that their great individual players could also perform as a team.

Polycentrism balances action and autonomy at multiple levels to create sustainable system behavior. Elinor Ostrom studied polycentrism and won a Nobel Prize (and let us know our lobster dinners can be sustainable); Vicente del Bosque practiced it and won a World Cup.

THE FAN WHO SCORED FOR WEST HAM, AND DIVESTMENT

Transparency

Lifelong West Ham United fan Steve Davies was drinking beer and smoking cigarettes with his wife and his friend Chunk as they heckled West Ham's underperforming center forward, Lee Chapman. Would you believe that Davies got to play and score for West Ham in that very same game, before the beer and cigarettes were out of his system?

Harry Redknapp, whom you might remember as Frank Lampard's uncle, was managing West Ham as they took on Oxford FC in that midsummer game in 1994. Redknapp heard Davies heckling from the sideline and asked the fan if he could do better. Davies, perhaps emboldened by the beer, responded that he could, in fact, do better. Then, before cooler heads prevailed, Redknapp escorted Davies to West Ham's locker room and got the fan a uniform and soccer shoes. Davies got dressed and, in the second half, Redknapp subbed the fan into the game—for Lee Chapman.

Davies had played some soccer and was in better shape than Chunk (which is not a nickname awarded to physical specimens). But Davies was completely out of place in the professional game, as you would expect of a fan pulled from the stands.

Even if this story ended with Davies floundering on the field, it would still be unbelievable. But that's not the end of the story.

Eventually, the ball found its way to Davies in a position where he had enough space and time to get off a shot. The ball caromed toward the goal and, miraculously, went in.

Imagine how Davies must have felt when that ball went over the goal line! Imagine being Redknapp. Imagine being Chapman.

The story is true, with some caveats. It was a preseason exhibition game, not a game that counted in the league standings. And it was a mismatch; West Ham were in England's top league at the time, whereas Oxford City played several leagues below. For West Ham, the game was more about practice than competition, a game to build cohesion and avoid injury. West Ham had brought only a few substitutes with them to Oxford and, after several injuries, Redknapp didn't have anyone on the bench to spell Chapman.

Davies's case is unique; heckling players from the stands won't get us invited to play in the game. However, as spectators, we get to choose whether to jeer or cheer because the information we need to do so is right there in plain sight.

Transparency means having access to the right information, in the right format, at the right time(s).

Transparency is vital in the campaign to divest from fossil-fuel companies. Students made me a believer in this movement as we discussed their well-reasoned proposal for our university (Clemson) to divest. Before talking with the students, I already knew that divestment generally means not financially supporting things you don't believe in—in this case the unsustainable burning of fossil fuels.

The students taught me more.

They explained how young activists had convinced the University of California to divest from companies doing business in apartheid-era

South Africa. Worldwide divestment followed and, eventually, the newly freed Nelson Mandela praised the young activists' contributions to the anti-apartheid struggle.

The students explained the massive financial influence of college and university endowments. Institutions have to spend just 5 percent of their endowments each year. Most take the rest and invest it to grow the endowment. Clemson's endowment is more than $500 million, and Harvard's is over $35 billion. In the United States alone, endowments add up to nearly half a trillion dollars—roughly equal to Argentina's gross domestic product.

And the students explained that, because of where most endowments are currently invested, even a university committed to renewable energy research is probably doing more to maintain the fossil-fuel status quo.[37]

Transparency ignited the fossil-fuel divestment campaign. Financial analysts studied data from fossil-fuel companies (and countries with nationalized fossil fuels) to estimate their reserves, which are fossil fuels the companies can access and are planning to burn. The study revealed that going through with these plans will release more than five times the amount of CO_2 scientists believe is safe to avoid the worst effects of climate change.[38] With this information, it's pretty hard not to support the divestment campaign.

Transparency can be enhanced by the format of information. If you had already heard about the plans to release five times as much CO_2 as we should, it probably wasn't directly from the analysts who crunched the numbers, but from a journalist who distilled the relevant findings. Author and activist Bill McKibben is a hero of the divestment campaign. There are more than half a million Google search results for his quote "We have five times as much oil and coal and gas on the books as climate scientists think is safe to burn."[39]

Just as format can boost transparency, so can timing of information. A one-time report raises awareness, but a persistent campaign needs

37. Imagine if Clemson decided to spend all of the required 5 percent of their endowment on renewable energy research. Even with that unlikely commitment, 95 percent of Clemson's endowment would still be invested. And most likely, the investment would be diversified across a stock market that is flush with fossil fuel companies.

38. The Carbon Tracker Initiative, which estimated the reserves and CO_2 from burning them, acknowledges that their estimates aren't perfect, but they are also not off by five times!

39. "Global Warming's Terrifying New Math" is Bill McKibben's article in *Rolling Stone* that reported on the Carbon Tracker Institute's findings.

5

updates to measure progress. The organization 350.org[40] promotes transparency by describing and tracking the divestment commitments of various institutions.

Despite the progress, transparency for divestment has a long way to go. Like most people, I would rather not financially support plans to burn five times as much carbon as is safe. But I ran up against a lack of transparency when trying to divest my family's savings.[41]

For one, our investments are mostly in indexed mutual funds, which are made up of hundreds of stocks chosen automatically to mimic the performance of the entire stock market. This includes fossil-fuel companies, and there is no option through our fund manager for an indexed

40. 350.org is named for the atmospheric parts per million of carbon we must stay below to avoid the worst effects of climate change. According to co2now.org, which provides transparency with updated measurements of atmospheric CO_2, we are at over four hundred parts per million as I write this. If we reduce CO_2 emissions, atmospheric levels can decrease.

41. My family's savings does not have the same clout as the endowments.

fund without fossil-fuel stocks.[42] This is basically the same excuse Clemson University is using for not divesting our endowment.

And what about non-fossil-fuel companies that rely on or contribute to unsustainable burning of these fuels, such as aluminum manufacturers and whoever is producing gas-guzzling Hummers these days? Even if I had time to look into every company in a mutual fund I was considering, I wouldn't have the transparency needed to make educated divestment decisions. What's needed is something like a carbon footprint for each company that is as prevalent and dynamic as the stock price itself. And that doesn't exist yet.

So, for now, I decided to do nothing.

Just kidding—I moved our savings into socially responsible index funds made up of stocks screened for social, human rights, and environmental criteria. We're not totally divested from fossil fuels, but we're much better than where we were before.

Don't count on transparency giving you a chance, like Steve Davies, to play for your favorite club. And transparency isn't good for everyone—it exposed the underwhelming performance of poor Lee Chapman, who ended up playing in ten scoreless games for West Ham that year before being sent away to a weaker team. But by engaging more than just the professionals in creative decision making for sustainability, transparency brings spectators onto the field.

Transparency means that people can see what is going on, whether in a soccer game or in their investment portfolio, which is a must if we expect to draw from their needed expertise to create sustainability. Even soccer fans who don't get invited to play for their team in the second half benefit from the Game's transparency.

"TAKE THE PISS" AND "LET THEM EAT CAKE"

Fixing Inequality

If you discuss soccer with Brits, or even with Americans who like to use phrases that originated in England, you may hear them say "take the

42. More and more mutual funds offer options that are divested from fossil fuels—check Google.

piss" or the seemingly less offensive (until you realize it's racist)[43] "take the mickey."

If Garrincha played for England instead of Brazil, his teammates would have said that "the bent-legged angel really took the piss" after he dribbled past his defender, and then turned around and did it again. Or, if our team is behind 5–0 in the pickup game and the other team is joking around, we might exhort our teammates with "C'mon, pick it up! They're taking the piss out of us." You get the idea; to take the piss is to remove someone's self-importance by mocking them.

Some players who take the piss in pickup soccer games are jerks. They do unnecessary tricks or play keep-away just to make their opponents look silly. Others taking the piss want the other team to continue playing, to keep the game going—or at least that's how I've rationalized my own mocking behavior.[44]

Everyone suffers when there is systematic inequality in pickup soccer. The losing team is miserable and quits, and the winning team doesn't have a game anymore.

To correct systematic inequality is to remove imbalances that are perpetuated by the system.

Correcting systematic inequality can also create more sustainable non-soccer systems. Unbalanced pickup games are trivial compared to systematic inequalities such as income gaps, lack of social mobility, and discrimination, often based on race or ethnicity. As in the pickup game, even the "winners" suffer—from complacency thanks to a rigged system, and from fear that the have-nots have finally been pushed past their threshold.

"Let them eat cake" was supposedly the response of France's Queen Marie Antoinette when she learned that peasants in her kingdom had no bread. There is no proof that Antoinette actually said those words, but the perception that she did insulted people who were unable to meet even their basic needs—and who certainly didn't have any cake.

The aloof perspectives and lavish spending of aristocrats like Antoinette led to the French Revolution at the end of the eighteenth century. During the revolution, the monarchy was dismantled and so were Antoinette and her husband, King Louis XVI, who were removed from power and executed by guillotine. It was a rapid fall for a royal couple

43. "Mickey" is a racist way to refer to someone from Ireland; the insulting and not even creative term refers to the many Irish last names that begin with "Mc" or "Mac."

44. Bear in mind that if you decide to take the piss to fix inequality, your opponent may respond by playing harder—or they may respond by trying to injure you.

who had been playing comfortably on the "winners" side of systematic inequality.

The French Revolution did spread fairer political beliefs: the resulting Declaration of the Rights of Man extended human rights to women and slaves (though not to future generations, as the Iroquois had already done). But it would have been nice to realize that progress without a brutal war. Wars are as unsustainable as it gets; when people are killing each other, the needs of the present are not being met.

We'd like to avoid wars, and there are better ways than taking the piss to avoid systematic inequality in pickup soccer. Someone who knows most of the players can divide them into teams, or there might be a more random[45] distribution based on shirt color. Perhaps even two captains could pick teams—and you can try the goalkeeper pick trick!

The objective when dividing into teams is to establish a fair and competitive game. It's no fun to play when your team is getting crushed. It's only slightly more enjoyable if your team is doing the crushing. The most gratifying game is closely matched, hard fought, brings out the best in both teams—and is one that your team eventually wins.

Even after trying to establish fair teams for the pickup game, there are balancing feedback loops in case the game becomes unbalanced. Players who show up late are added to the team that is behind. Maybe a couple of the best players are switched from the dominant team to the losing one. The score might be wiped clean, providing a fresh start for the team that was behind and new motivation for the one that was winning. Or a hopeful player on a losing team may yell out "Next goal wins!"—and occasionally those on the team that is winning may even agree.

Systematic inequality is not sustainable for pickup soccer or for royalty headed to the guillotine. No one is truly insulated. Even when we are safe from an uprising, inequality creates a false sense of security, which breeds complacency, which prevents progress. A level playing field helps everyone get better.

Whatever the system, inequality hurts both the haves and the have-nots. So, even when our morals don't guide us toward equality, the desire for a more competitive soccer game should.

Inequality leaves systems vulnerable to inevitable shocks and limits opportunity for the have-nots—while, in the haves, it can foster a cake-eating

45. Experienced pickup players coordinate with friends to make sure they all wear the same color of shirt to the field.

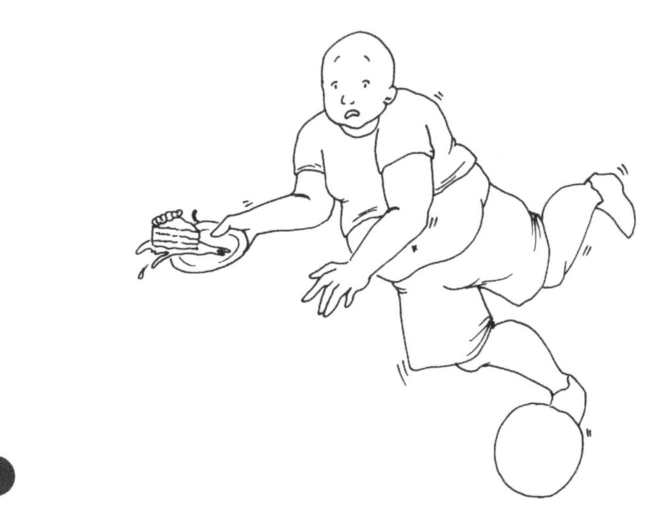

lack of fitness. Balancing feedback loops can correct systematic inequality, as we see in a new story about an old revolution and in those who tease others in pickup soccer.

BREAKAWAYS, PASS-BACKS, AND MY REPURPOSED OFFICE

Elegance

The breakaway penalty shot was a novelty that, in the late 1990s, found its way into Major League Soccer (MLS), the top level of professional soccer in the United States.[46]

Penalty shots are the punishment when the noble but clumsy defender commits a foul in their penalty area, or when an unethical but clever attacker tricks the referee into thinking a foul was committed. On penalty shots all over the world, the shooter has one kick to try to score against the goalie from twelve yards away. Penalty shots are scored more than three-quarters of the time.

MLS's breakaway penalty shot was meant to enliven the time-tested stationary penalty shot. Instead of a static kick from twelve yards, the Americanized version had players starting forty yards away from the goal with five seconds to do whatever they wanted to try to score against only the goalkeeper, who was also allowed to move anywhere. Typically,

46. MLS was officially founded in 1993, in part to justify the United States' hosting of the 1994 World Cup and in part to capitalize on the resulting marketing boost. The first MLS game took place in 1996.

the goalkeeper (to cut down the shooting angles) and shooter (to get a closer shot) sprinted directly toward each other as soon as the five-second countdown began.

The other MLS twist was to settle tied games with shootouts in which five players from each team attempted breakaway penalty shots. Ties are an accepted part of the Game in the world's best leagues. A tie on the road against the league leaders can feel like a win. Penalty shootouts (the regular kind) are used only in playoff scenarios, when one team must be chosen to advance to the next game. And even in these cases, the penalty shootout is used only if extra periods do not break the tie.[47]

The breakaway shootout and its use to settle tied games were responses to a prevailing notion that soccer would never catch on in the United States unless the Game was made more exciting. Americans, the thought was, need to see cheerleaders, time-outs, commercials, smoke-filled player introductions, and big balls dropping from the rafters.

But soccer works because it doesn't have all that fluff. It works because the Game is simple. A father who grew up playing baseball can understand what his soccer-playing daughter is trying to do (get the ball in the other team's goal) and what she's not allowed to do (use her hands or smash into other players). Details like the offsides rule (which we'll get to later) take longer to appreciate, but Dad can enjoy watching his daughter's game right away.

Soccer also works because it is subtractive. Some of the best things about the Game are what it *doesn't* have. There are no restrictions on where players can go on the field. There are no time-outs,[48] fewer referees, and fewer substitutions than in other sports. MLS's breakaways didn't subtract for beautiful simplicity; they added for unnecessary complexity.

And soccer works because it is surprising. Goals completely change the status of a game and come in an instant, sometimes from a team that seems to be getting pummeled. Lower scoring means that goals are less

47. Usually, two fifteen-minute periods of extra time are played to try to settle ties before penalties. But the rules during my final year in college called for four fifteen-minute overtime periods, the last two being "golden goal," where the game was over if one team scored. We won our league championship with two straight golden goal games. Then, after 150 minutes of tied soccer in the first game of the national tournament, we lost on (non-breakaway) penalty kicks.

48. Even without commercial interruptions, professional soccer clubs still find creative ways to advertise during the games—in particular by selling the space on players' jerseys, which works on me. I'm way more likely to buy the mobile phone I see on Zidane's jersey than the one pitched by an actor in a commercial that is interrupting what I really want to be watching.

frequent, but it also makes the goals that much more precious. In basketball, you score and it's expected, so you run back on defense as fast as possible. In soccer, you score and it's "almost as good as sex," at least according to my coach's pregame speech. MLS's breakaways did not add surprise because they were just as arbitrary as penalty kicks.

Soccer is elegant, which means simple, subtractive (in a good way), and surprising. And elegant systems tend to be more sustainable.

Pursuit of elegance inspired me to abandon my office at work, which is one of the most sustainable behavioral changes I've ever made.

Let me provide some background. My department is raising money for a new building, and there's no question we can improve on what we have: lizards occupy light fixtures, and desks from the 1950s remain bolted to the floor. But I'm not sure we need to add space. We have more students than we used to, but these students have different needs and opportunities, like online courses and industry co-ops, that change how we (should) use the space.

Questioning the need to add space, I realized that I could use my office to test how student needs align with the space we already have. At the time, I was using my office for about six hours in a typical week. The rest of the time it sat unused.[49] I realized that the missed opportunity was not how my office was being used, but when.

That epiphany was the beginning of the end for my office as a private dwelling. I took out the big desk, extra monitor, printer, files, and the books I had just to show how many books I had. My office became a blank canvas for new uses.

The space now has a small café table and a few stools for student meetings, which are what I typically used the office for to begin with. Students use the office when I'm not there. And as a result, within a year of the change, the space was being used about three times more than when it was mine alone.

Before building new, it makes sense to think about whether we can be smarter with what we already have. A new building makes for a nice ribbon-cutting ceremony but also consumes loads of resources. A new building may add space, but will it meet student needs? After all, the breakaway shootouts eliminated ties, but they didn't guarantee entertaining soccer.

In 1992, several years before MLS tried breakaway shootouts, a more subtle rule change was introduced worldwide: goalkeepers were no longer permitted to use their hands on passes from their

49. To be clear, I work plenty, it's just that nearly all of my writing, teaching, research, and meetings happen somewhere other than my office.

teammates.[50] Watch a game from before the rules changed, say the 1990 World Cup, and you will see long backward passes to the goalkeeper—who, at the time, was allowed to pick the ball up.

After the back pass, a goalkeeper who wanted to waste time to protect a lead or a tie could roll the ball to a teammate, receive a pass back, and then pick up the ball again, and on and on. This scheme was one reason that the 1990 World Cup had fewer goals per game than any other (despite the contributions from the Cameroonian Milla and the Colombian El Loco).

Even worse than the lack of goals was the overly cautious play as most teams decided not to pressure their opponents, who could escape by passing to their goalkeeper. When goalkeepers were banned from using their hands on passes from teammates, it eliminated a safety valve, and teams began applying pressure with their attacking players. The pass-back rule change remains in place.

If you're among the millions who now watch MLS games,[51] you know that breakaway shootouts are extinct.[52] They died because they added complexity, and the more complicated something is, the more likely it is to fail in unexpected ways. By contrast, the pass-back rule found a place in the Game by mirroring the simple, surprising, and subtractive elegance of soccer.

Elegant solutions may seem simple after the fact, but coming up with them is not easy. My repurposed office required a willingness to break from the status quo (and a cooperative department chair). Soccer's pass-back rule came from identifying the root problem and considering a range of alternatives.

Or, as the Dutch legend Johan Cruyff put it: "Simple football is the most beautiful. But playing simple football is the hardest thing."

50. The International Football Association Board (IFAB) maintains the rules of soccer that govern play around the world. FIFA competitions like the World Cup use these rules. National and regional federations will occasionally experiment with rule changes like the breakaway shootout.

51. It turns out that Americans enjoy pretty much the same version of soccer as people all over the world. Attendance is higher, on average, at Major League Soccer games than at National Basketball Association or National Hockey League games. On TV, Americans are increasingly enjoying forty-five-minute halves of unpredictable drama that are uninterrupted by commercials for erectile dysfunction pills. TV ratings were higher for U.S. national team games during the 2014 World Cup than they were for baseball's World Series that year.

52. I was pro–breakaway shootout, mostly because I supported any rule change that gave me an extra opportunity to score. In case you're wondering, my strategy was to dribble as fast as I could at the goalkeeper, pretend to shoot as hard as I could right at him, and—when he (hopefully) flinched—dribble past to score on an open net.

Elegance, whether for soccer or sustainability, is this high-performance simplicity.

Elegance isn't easy, but the good news is that high-performance simplicity brings with it both function and beauty, which is why Pablo Picasso defined art as the elimination of the unnecessary. And it's why simple soccer is known as the beautiful Game.

Elegant systems tend to be more sustainable. That's why we don't see breakaway shootouts anymore and why we should consider rearranging the furniture before lobbying for a new building to work in.

Review: Creating

Biomimicry means asking ourselves what in the natural world already does what we are trying to do, and then learning from that example to make our approach more sustainable.

Closing loops means eliminating waste and ending reliance on inputs from outside the system. (It's also a type of biomimicry.)

Adaptability is being able to adjust to disturbances by modifying behavior—without any outside intervention. Diverse systems are more adaptable.

Polycentrism is a balance of management and autonomy at multiple levels to sustainably manage a shared resource.

Transparency means having access to the right information, in the right format, at the right time(s).

To correct systematic inequality is to remove imbalances that are perpetuated by the system.

Soccer is elegant, which means simple, subtractive (in a good way), and surprising. And elegant systems tend to be more sustainable.

7

The Endless Quest

There is no magical fix; pursuing sustainability is an endless quest of constant effort at the limits of our abilities, and that's where the fun is.

MAPUTO, MOZAMBIQUE, AND THE WORLD

Overview: The Endless Quest

The more I study sustainability, the more I appreciate that pursuing it is an endless quest. Sustainability won't be handed down by supposed superiors; it will be a new reality we create for ourselves.

Climate change alone requires constant vigilance. And climate change intensifies all the other sustainability issues we've discussed: from growing and urbanizing populations, to ecosystems scarred by human activity, to economies that rely on abuse of natural and human resources.

Responding to these issues isn't easy. It demands thinking that is global, multidisciplinary, and multigenerational. And, in many cases, we face entrenched financial interests.[1] But that's all right with me, because endless quests thrill me and I'm bored when I sit back and let it all be.

Plus, whether or not we are thrilled by endless quests, we are all on them already. The quest is the essence of Ralph Waldo Emerson's often-repeated observation that "life is a journey, not a destination." Nelson Mandela's quest was why he said, "I dare not linger, for my long walk

1. In an essay for *The Nation,* "The New Abolitionism," Chris Hayes argues convincingly that the abolition of slavery provides a template for the climate justice movement because embedded interests will have to "say goodbye to trillions of dollars of wealth."

is not ended." I like how the soccer legend Valeriy Lobanovskyi[2] put it: as soon as his Dynamo Kiev club team won their first-ever Soviet league championship, Lobanovskyi immediately reflected that "a realized dream ceases to be a dream."

We get nowhere on our quest if all we do is study sustainability. Learning every last detail about a city like Maputo is impossible, and trying to do so would delay much-needed action. Eventually we have to do something, even though we're not sure what the future holds. This dilemma is why we need a *vision,* our dream for what the more sustainable future will look like.

Let's say a vision for Maputo is that everyone, rich or poor, has access to basic needs and equal opportunity to advance socially and economically. That vision then guides long-term planning strategy, such as whether to privatize the city's water supply. The vision also informs rapid reactions to sudden events, such as new flooding in a previously dry slum. Whether for long-term water planning or short-term flood response, people can prioritize action that moves Maputo toward the shared vision for basic needs and equal opportunity.

One way to pursue a vision is by changing rules, and rules changed drastically when Mozambique went from a Portuguese colony to an independent nation in 1975. Although they can help, new rules are not a cure-all. When Mozambique gained independence, the mass exodus of Portuguese interrupted basic services and triggered fighting over the roles left behind (imagine the chaos where you live if public services and the people who provide them were gone tomorrow). Not only were the systematic inequalities still present in Mozambique, but roughly one million people died and millions more were displaced during the fifteen-year civil war that followed independence.

Fortunately, changing rules is just one of many ways to pursue our visions. We can find and push *leverage points,* which are places where a small shift can produce big changes. There are leverage points everywhere in a system—in elements, flows, stocks, feedback loops, and purposes. And we can push leverage points without formal rules. Mozam-

2. After his playing days, Lobanovskyi managed Dynamo Kiev, and his scientific (and systems-thinking) approach to the Game brought the club unmatched success: Kiev won eight Soviet league titles and two European Championships in Lobanovskyi's initial fifteen-year stint as manager. When he later returned to a then struggling Kiev club, they went on to win five straight league championships and make the semifinals of the European Championships. Dynamo Kiev now plays in a stadium named for Lobanovskyi.

bique's civil war refugees pushed a powerful leverage point for equality by coming home: as more and more refugees returned, it became safer for others to do so. Five million people have returned to their homes in this reinforcing feedback loop, which is the largest repatriation ever in Sub-Saharan Africa. When people got home, they could finally seize the new social and economic opportunities that came with independence.

The greatest leverage points are *mindsets:* the values, ideas, and assumptions from which our systems arise. Mozambique's civil war raged on until enough mindsets shifted from thinking that warring armies would resolve conflict to thinking that political parties were worth a try.

On the quest for sustainability, we envision the future we want to live in and push the leverage points we think will get us there. The quest is endless because we then reflect and do it all over again.

CRUYFF AND COSTA RICAN CARBON NEUTRALITY

Visioning

Few can rival Johan Cruyff's imprint on the Game as a player, manager (he didn't just suck lollipops), technical director, and pontificator.[3] In all of his roles, Cruyff is synonymous with soccer vision.

Unquestionably the best player ever to wear the number 14,[4] Cruyff won soccer's top individual honor, the Ballon d'Or,[5] three different times in the early 1970s. He also led his tiny club team, Ajax of Amsterdam, to three straight European Cup Championships. One reason Cruyff was such an effective player is that he saw possibilities that were several passes ahead, a type of short-term visioning that only elite playmakers develop.

3. My favorite Cruyff pontification is when he said to pestering reporters, "If I wanted you to understand it, I would have explained it better." Unlike Cruyff, I was never good enough to find reporters pestering.

4. Cruyff wore the number 14. I allowed people to think that Cruyff was the reason I wore number 14 so I wouldn't have to explain that I wore the number to remind me of the league my dad and his friend started just so their kids could play soccer. I was number 14 in each of my first two years on the light blue team because my size put me in the large t-shirt category and because, as the coach's son, I got whatever number no one else wanted.

5. The Ballon d'Or ("Golden Ball") award began in 1956 and was voted on by journalists. In 1995, non-Europeans became eligible for this award, but only if they played for a club in Europe. In 2010, the Ballon d'Or was merged into the existing FIFA award for a true World Player of the Year.

Playing for his country, Cruyff helped Holland qualify for their first-ever World Cup in 1974. They made it all the way to the final game, where they faced West Germany. Holland took the opening kickoff and made thirteen consecutive passes before Cruyff was fouled in the penalty box. His teammate scored the penalty kick, and Holland were ahead 1–0 before the West Germans had even touched the ball. Despite going on to lose 2–1, that Dutch team was the story of the tournament. They introduced the world to an artistic playing style that freed field players from unnecessary positional rigidity (similar to what El Loco Higuita would later do for goalkeepers).

This new free-flowing style of play earned the name "Total Football." Soccer connoisseurs endlessly analyze the intricacies of Total Football, but the vision is quite simple: any player on the field is capable of filling the role of any other player.[6] Or, as Cruyff put it, "attackers could play as defenders and defenders as attackers. Everyone could play everywhere." That vision guided everything about Total Football, including the emphasis on ball possession,[7] spacing,[8] and attacking play.[9]

Following the 1974 World Cup, Cruyff moved from Ajax to the Spanish club Barcelona. While playing at Barcelona, Cruyff took the first steps in his vision to instill the Total Football philosophy in all of that club's players. He taught the managers for Barcelona's youth programs so that the future professionals would develop the adaptable skills best suited to Total Football.[10]

After winding down his playing days at smaller clubs, Cruyff later returned to Barcelona as a lollipop-sucking manager. The players who were molded by Cruyff's vision on their youth teams were now professionals contributing to Cruyff's managerial successes, including Barcelona's

6. As with any tactical innovation, Total Football did not arise out of thin air. The Argentine player and manager Carlos Peucelle described positionless soccer when he referred to his dominant 1940s River Plate club as having a goalkeeper and ten field players (as opposed to a certain number of forwards, midfielders, and defenders).

7. Cruyff on possession: "There is only one ball, so you need to have it."

8. Cruyff on spacing: "If you have the ball you must make the field as big as possible, and if you don't have the ball you must make it as small as possible."

9. Cruyff on attacking: "For me attack is the best defense."

10. This Total Football principle of adaptable players has been adopted all over the world. Italian manager Arrigo Sacchi calls this trait in players "universality." Portuguese visionary José Mourinho put it in his characteristically other-deprecating, yet accurate, way: "I can't believe that in England they don't teach young players to be multi-functional."

triumph in the 1992 European Cup. Cruyff had made his vision real, and it wouldn't be the last time he did so.

A vision is our dream for how we want things to be in the future. Our visions guide us when we encounter unforeseen decisions and obstacles on our sustainability quests.

In 2007, twenty-five years after Cruyff managed Barcelona to victory in the European Cup, Costa Rica announced a vision to become the first carbon-neutral country by 2021. Costa Rica's 2021 vision affects national-level decisions about budgets, laws, and incentives. And, as a well-publicized national goal, the vision also shapes individual decisions—for example, whether or not to install solar panels on a home, or whether to drive alone or carpool.

To reach their 2021 vision, Costa Rica will need to slash energy consumption and meet the remaining demand with renewable resources such as solar, wind, and biofuels that are consumed more slowly than the regeneration rate. Achieving carbon neutrality won't be easy. It requires a complete shift in culture. But Costa Rica has a history of visionary change. They got rid of their military in 1948, spending that part of the national budget on education and culture instead.[11]

The bold carbon neutrality vision has already had spillover benefits.[12] In 2012, Costa Rica was named the greenest country in the world after ranking first in the Happy Planet Index[13] (an index that would probably not exist without Bhutan's earlier vision for "gross national happiness"). With the publicity from such accolades, Costa Rica is hosting more and more tourists from around the world, including my brother and his wife, who took their international honeymoon there.

Just as Bhutan did with gross national happiness and Costa Rica is doing with carbon neutrality, Cruyff used a vision to break from the status quo over a long time span. Cruyff may have won the European Cup as manager in 1994, but even more impressive results were brewing.

11. Since getting rid of their military, Costa Rica has not had a civil war and has not been taken over.

12. OK, the following spillover benefit is probably not really correlated with Costa Rica's 2021 vision for carbon neutrality, but . . . they were the surprise team of the 2014 World Cup, making it all the way to the quarterfinals.

13. The Happy Planet Index is a reputable measure of human well-being that is weighted so that higher scores are earned by nations with lower ecological footprints. I wish they awarded World Cup hosting duties based on this metric, but that's not how it works. Among the lowest-ranking countries in 2012's Happy Planet Index was Qatar, our host for the 2022 World Cup.

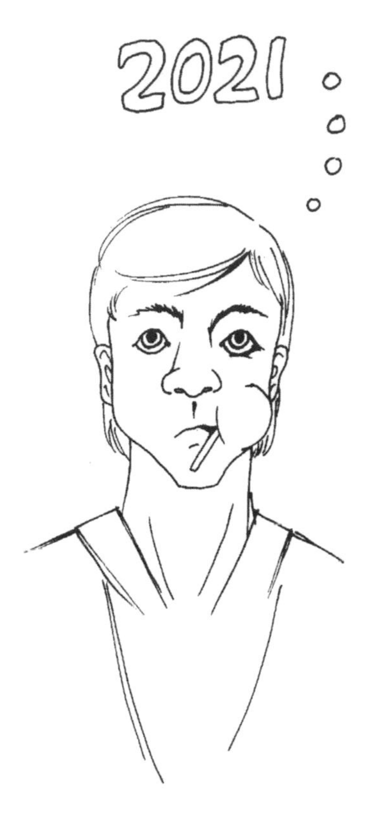

Three decades after Cruyff began his vision for youth development came the Barcelona team that staked their claim as the best of all time, the team that could be derailed only by Icelandic volcanoes. Of the eleven players who started for Barcelona in their 2009 European Cup final victory, eight were alumni of Cruyff's youth system. In that year's World Player of the Year voting, three of the top five finishers came through Cruyff's visionary system, including the winner, Lionel Messi.

Cruyff's vision shaped Spanish soccer too. The Del Bosque–managed (and polycentric) Spanish national team that lifted the 2010 World Cup trophy started six players from Barcelona's youth system in the final. One of them, Andrés Iniesta, scored the winning goal in extra time against, ironically, Holland.[14]

14. Doubling the irony, in the 2010 World Cup final against Spain, the Dutch team played essentially the opposite of Total Football. As Cruyff put it: "This ugly, vulgar, hard, hermetic, hardly eye-catching, hardly football style, yes it served the Dutch to unsettle

The more clearly we can envision the future we want, the more likely it will happen. A clear vision didn't let Cruyff predict the future of soccer, but it did help him create it. Similarly, Costa Rica's vision for carbon neutrality won't prevent surprises on their quest, but it will help them turn these would-be obstacles into stepping stones. And we'll all be better off if Costa Rica's vision for carbon neutrality is as successful as Cruyff's vision for Barcelona.

With a clear vision for a more sustainable future, we can compare the merits of possible actions. The Dutch master Johan Cruyff exemplifies soccer vision, and Costa Ricans show vision in response to climate change.

OFFSIDES AND REFRIGERATORS

Rules

It's soccer limbo. The ball is in the back of the net, but the goal-scorer has not fully committed to the (premeditated) celebration, and the team that was scored on has one last chance. Is the linesman's flag raised to signal offsides and negate the goal?

As a player, I constantly tempted the offsides rule, trying to receive or deliver passes behind the last defender. While I don't remember ever being legitimately offsides, I will never forget beautiful goals disallowed by linesmen too slow of foot to be in the right position or too slow of mind to know that I was allowed to be past the last defender *as soon as the ball was passed.*[15]

Whether you're a player or a fan, it's deflating when a goal for your team is disallowed for offsides. Remember Steve Davies, the West Ham fan who got called into a game? His miraculous goal, the one I told you about earlier, was called back because he was offsides. See how disappointing offsides can be!

Spain. If with this they got satisfaction, fine, but they ended up losing. They were playing anti-football."

15. The italics simulate the "encouraging" tone my dad would use to remind offending linesmen of this key stipulation in the offsides rule. In fairness to linesmen, calling offsides requires seeing and comparing three separate moving objects at once—which, according to eye specialists, is impossible.

Soccer's rule-makers have amended the offsides rule, mostly in attempts to reward entertaining attacking play.[16] It used to be that at least three defenders had to be between the offensive player receiving a pass and the goal. In 1925, that rule was changed so that only two defenders were required (one defender is almost always the goalie—except when you have a crazy one like Higuita running way up the field). After the rule change in 1925, goal scoring, an imperfect but vital measure of soccer's entertainment value, immediately went up by more than a goal per game.[17]

When they work as intended, rules can be dynamic disturbances that improve systems and move us toward our visions.

Full disclosure: I'm not a rule-lover. Like most professors and attacking players, I despise the restrictions. And I know we can't fully control systems with rules any more than we can predict who will win the next World Cup. Yet even I appreciate that, in certain cases, rules can advance our sustainability quests without restricting freedom.

Consider the case of refrigerators, for example. Refrigerators in the United States used more and more energy per unit every year from their invention until the 1970s, when the first efficiency rules were established. The new rules didn't just stop the upward trend; they reversed it. A refrigerator built in 2012 uses about one-quarter the amount of energy per unit as a refrigerator built forty years earlier, before the efficiency rules were put in place.

The average family saves about $150 in annual energy costs because of more efficient refrigerators.[18] Altogether, the efficiency rules have saved us from burning roughly seven quads[19] of energy (estimates vary; seven is on the low side). To produce that amount of energy would take

16. Why not just eliminate the offsides rule? In the middle of their 1972 season, the North American Soccer League tried a rule whereby players could only be offsides eighteen yards or less from the goal line. The next three matches produced just three goals—total. And the average number of goals per game fell for the remainder of that season. Defenders simply dropped farther back, making it even harder for attackers to get behind them.

17. Goals per game, which had dipped to 2.58, increased to 3.69 the year after the offsides rule was changed (see www.theguardian.com/sport/blog/2010/apr/13/the-question-why-is-offside-law-genius).

18. For more about refrigerator standards, this is a good overview: http://energy.gov/articles/proof-pudding-how-refrigerator-standards-have-saved-consumers-billions.

19. One "quad" is a quadrillion British thermal units, which is roughly the amount of energy in 45 million tons of coal, which is about 920 million cubic feet.

about three hundred million tons of coal,[20] enough to make a pile that is as tall as my false-brooding brother (two meters), five times wider than the neighborhood-dividing Cross Bronx Expressway (eighteen meters), and as long as Barcelona's volcano-induced bus ride to Milan (one thousand kilometers).

As that massive pile of coal attests, rules can help. But there are drawbacks. Rules require enforcement. When hard-to-enforce rules prohibited the sale of alcohol in the United States in the 1920s, organized crime swept in to meet demand (I'm not aware of any bootleg refrigerators).

Rebound effects also can limit the effectiveness of rules. That $150 American families save each year thanks to refrigerator rules means that we have more to spend on other things, perhaps even another refrigerator. Still, when it comes to energy, the rebound effect typically negates only a small part of the initial savings—unless the savings are needed to meet basic needs.

The rebound effect eats away at the goal-scoring gains from the 1925 offsides rule change. Every World Cup since 1960 has averaged fewer than three goals per game, and recent per game averages are back down near pre-1925 levels. Just as people who spend less on energy have more to spend on stuff that uses energy, defenses adapt to new offsides rules. Before 1925, most teams played with only two defenders. Now, in part to cover for the last defender exposed by the revised offsides rule, teams often line up with five dedicated defenders.

And again, rules require enforcement. Rules are why soccer has a center referee, two linesmen, and sometimes even a fourth official, like the one who tattled on Zidane for his head-butt. You know by now that referees are the reason for most games I lost and goals I missed.[21] But we need referees because rules don't work when they are not enforced.

Our endless quests will come with new rules, which will require enforcement and may lead to rebound effects. But well-crafted rules can unleash creativity—not to beat rules, but to pursue the underlying

20. Seven quads is around 6.44 billion cubic feet or 182 million cubic meters, which comes out to that big pile stretching from Milan to Barcelona. These are just estimates of course, but they are conservative—and the bottom line is that refrigerators have saved us from burning a biiiig pile of coal.

21. The soccer sage Eduardo Galeano says of blaming referees: "We would have to invent them if they didn't already exist."

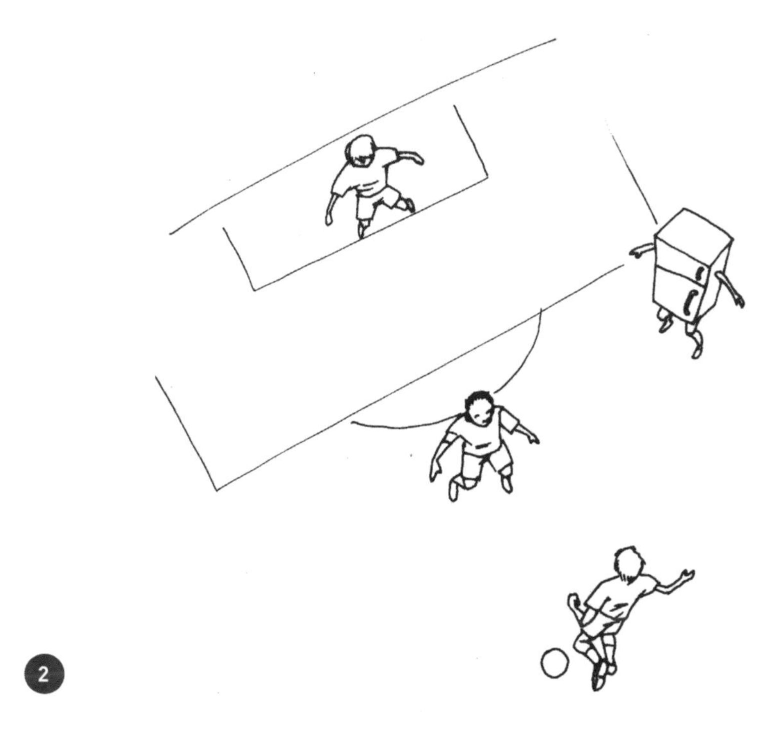

purpose. And when that is the case, rules can slash energy use—and increase goal scoring.

Rules can stimulate or stifle the novelty, creativity, and flexibility needed for sustainability. That's why some changes to the offsides rule have endured and why refrigerator standards in the United States have saved us from burning a highway of coal.

BERTHA, DILMA, AND MARTA

Leverage Points

You now know that Brazil produces transcendent soccer players, and there's one more we all need to appreciate. The most dominant Brazilian player since the turn of the century was named World Player of the Year five consecutive times, led Brazil to the World Cup final, was top scorer at that World Cup, became the all-time leader in World Cup goals, and averaged a goal per game—a record that hadn't been achieved since the legendary Pelé's era.

Yet the on-field soccer exploits of Marta Vieira da Silva pale in comparison with what she has done for gender equality in Brazil and beyond.

Sure, Brazilian women had rights before Marta was scoring goals. They gained the right to vote in the 1930s, and the same rights as men as part of Brazil's 1988 constitution. These legal rights advanced equality in health treatment and access to education.

Yet, in terms of reproductive rights, empowerment, and labor-force participation, Brazil does worse than neighboring countries (and soccer rivals) like Argentina and Uruguay.[22] The systematic gender inequality is even worse in the rural northeast, where children ate dirt, and where Marta grew up.

In Brazil, it was illegal for women to play soccer as recently as the 1980s. Marta's teammate Pretinha represented Brazil in four World Cups. But to do so, she had to overcome beatings for playing soccer from her brothers and her mother.

Even when they are not physically assaulted, female players in Brazil are pressured to project an image in line with social expectations. The pressure starts at the top. In 2004, FIFA's dinosaur president Sepp Blatter lectured that female players should "wear tighter shorts and low-cut shirts to create a more female aesthetic."

Young girls are steered away from soccer even by well-meaning parents, perhaps to prevent physical injury, though more likely out of fear their daughters will be ostracized. Some fan clubs in Brazil still exclude women, even though more Brazilian women than men watched the 2010 World Cup.[23] The notion that soccer is a man's game has stubbornly persisted.

But Marta has permanently cracked this antiquated notion. She has made it common for women to enjoy soccer the way men have for centuries. She has improved life for millions of Brazilian women, not through rules and laws but through her pioneering example. With her flair and function, Marta pushes change-making leverage points by showing more women they can enjoy soccer, and by smashing stereotypes among those who might otherwise stand in their way.

22. These gender equality measures are part of the United Nations' Gender Inequality Index (http://hdr.undp.org/en/content/gender-inequality-index-gii).

23. In the United States, women made up only about one-third of the 2010 World Cup television audience (see www.fifa.com/mm/document/affederation/tv/01/47/32/73/2010fifaworldcupsouthafricatvaudiencereport.pdf).

Leverage points[24] *are places where a small shift in one thing can produce big changes in everything. We can push leverage points in all the parts of a system: the elements, flows, stocks, purposes, and mindsets.*

Like the soccer heroines Marta and Pretinha, Bertha and Dilma deserve one-name recognition as pioneers of gender equality in Brazil.

Bertha Maria Julia Lutz was a scientist who studied poison dart frogs. If you love amphibians, as my dad and my young neighbor Lewis do, you will be impressed that Bertha has a frog, "Lutz's Rapids Frog" (*Paratelmatobius lutzii*), named for her.

While studying amphibians at college in France, Bertha also learned about feminist movements in Europe and the United States. Later, in 1922, Bertha traveled to the United States to attend the Pan American Conference of Women in Baltimore. There she learned from the Uruguayan Paulina Luisi and the American Carrie Chapman Catt, who were leading the women's rights struggles in their respective countries.

Bertha took her knowledge and networks back to Brazil and used them to start the Brazilian Federation for the Advancement of Women. By 1933 the group had enough political power for Bertha to get chosen to draft the first page of Brazil's new constitution.[25] She wrote in voting rights for women, as well as rights to hold government office and to earn equal pay. Bertha even gave preference to women in government jobs dealing with the home, motherhood, children, and working conditions for women.

The constitution went into effect in 1934 and was a landmark change toward gender equality. At that time, Brazil was one of only six countries where women had the right to vote.[26]

Bertha's work pushed a leverage point by changing allowable actions of system *elements*—women could now vote! The change continues to advance equality in Brazilian society.

To see how, let's meet our third Brazilian heroine, Dilma Vana Rousseff. When she was just a teenager, Dilma fought as an urban guerrilla against Brazil's military dictatorship. For trying to improve her country, Dilma eventually spent two years in prison. She was being tortured

24. One of my academic heroes, the late sustainable systems expert Donella Meadows, popularized the phrase "leverage points" through her writing, speaking, and teaching.

25. The new government was Getúlio Vargas's military regime, which had taken control of the country in 1930.

26. The military dictator Vargas allowed suffrage to relieve political pressure, but, in general, his regime repressed the budding women's movement.

there in 1970 as the Brazilian soccer legend Pelé won the last of his World Cups.

Ten World Cups later, in 2010, Dilma was elected as the first female president of Brazil.

Dilma's election was a breakthrough because there is a big step between being "allowed" to do something and that thing actually being done. Women have always been allowed to run for president in the United States. Yet after fifty-six elections and counting, the United States has never elected a female president. The chance that the streak of male presidents is random, and not due to a systematic inequality, is around 1 in 72,000,000,000,000,000 (quadrillion).[27]

Our three heroines push leverage points in system elements, and they also change the world by changing *flows*.

When Bertha drafted the constitution that gave women the right to vote, that led to another flow of information: the feedback of women via their election of public officials. A group that can vote is more likely to have their voices heard and their needs addressed.

Dilma pushes leverage points in change-making information flows by being a role model. Women see Dilma on television; they hear her speeches and read about her. Dilma's example is a flow of information that gives these women a path and a vision, just as Frank Lampard's father did for him and as my dad did for me.

Bertha, Dilma, and Marta all chip away at huge *stocks* of built-up resistance. In Brazil, equality of all types, in particular for women, must overcome the male-controlled culture that colonized the country.[28] Colonists "protected" natives from warring tribes in exchange for gold and labor, which has left a legacy of economic dependence among women.

The built-up stock of cultural norms oppressing women did not go away with the passage of any single law. When Bertha gained the right for women to vote, it was only for those who were literate, and most were not. But in her ensuing political career, Bertha used legislation to advance women's rights to work, equal wages, and maternity leave.

And while Dilma broke the glass ceiling for future female presidents in Brazil, cultural barriers remain. Women make up less than a

27. Or, put another way, the chance of randomly selecting fifty-six consecutive male presidents would be like Paul the Octopus (if he were still alive) correctly picking the winner of fifty-six soccer games in a row. If that comparison means nothing to you, just trust me that there is systematic inequality in the selection of U.S. presidents.

28. Sexist people are slowly dying off, some faster than others—the dictator Vargas committed suicide immediately after losing power.

tenth of Brazil's Parliament, despite making up at least half of all Brazilians.[29]

Stocks like the cultural norms in Brazil are daunting. But flows in the right direction, like those spurred by Bertha, Dilma, and Marta, will eventually overcome these stocks and move the system past a threshold.

By helping women gain the right to vote, Bertha started a virtuous reinforcing *feedback loop*. With the vote, women in Brazil can act like a pickup soccer game in which the individual players self-organize into a functioning team. Women in Brazil can create new structures, add new information flows, and make new rules. Even more women gain the right to vote, and the feedback loop goes on.

Dilma has also created a reinforcing feedback loop. Her personal approval rating soared close to 80 percent in 2013 because of popular measures such as exempting federal tax on staple foods.[30] Despite a turbulent second term after her reelection in 2014, she was ranked as

29. Women remain overrepresented in one area: of every ten poor people, seven are women.

30. Dilma also removed federal tax from energy bills, which may not be the best policy for climate change. And, though she has supported same-sex civil union and criminalization of homophobia, Dilma does oppose same-sex marriage. Her official stance is that "marriage is a religious issue. I, as an individual, would never say what a religion should do or not. We have to respect them." "Them" being religions, not people.

the thirty-first most powerful person in the world, which I mention mainly because that was thirty-nine spots ahead of the sexist FIFA president, Sepp Blatter. Dilma's power brings more leadership opportunities for women, which will result in more women serving as inspirational examples to other women, and so on.

Finally, some of the most influential leverage points are those that affect the *purpose* of the system, because changing purposes affects all the other parts of the system: elements, flows, stocks, and feedback loops.

Bertha changed the purpose of Brazilian government: from representing literate men to representing literate people.

Dilma shifts the purpose of the system with her pledge that Brazil will continue to grow in social mobility and inclusion—two areas where infinite growth is actually possible on a finite planet.

And Marta has changed the purpose of Brazil's soccer system and the world that connects to it. She is helping more people realize that having a Y chromosome is not a requirement for enjoying soccer.

Leverage points are places where a small shift produces big changes. Bertha, Dilma, and Marta have all pushed leverage points to correct systematic inequality for a more sustainable Brazil.

THE GREATEST LEVERAGE POINT AND MARTA CONTINUED

Mindsets

The greatest leverage point is one we can appreciate by remembering our shared stories.

Soccer stories showed us that Benfica and Portugal won not only because the Black Panther scored but also because the Sacred Monster passed and defended. We know that Brazil didn't give away the 1950 World Cup final—Uruguay took it. And you know why I still blame a volcano for interrupting Barcelona's three-peat, even though "The Tractor" Zanetti and Inter Milan deserve most of the credit.

The goalkeeper El Loco broke through the implied boundary of the penalty box. The manager Del Bosque showed that leaders need not be dictators. The genius Panenka realized that it's not just the ball that has inertia on penalty kicks—the goalkeeper has it too. Holland's Total Football was a revolution in player tactics from rigid positions to flexible patterns. These beautiful changes to the Game all sprung from mindset changes.

Mindsets are the values, ideas, and shared assumptions that are the foundations of society (and soccer).

The mindset from which a system arises is the greatest leverage point we have;[31] it's like the catapult of leverage points. Sure, systems are hard to change when values, ideas, and assumptions are deeply rooted. But if the prevailing mindset is flawed, we don't need resources or rules to push the mindset's leverage point; we just need people with open minds. As new beliefs and norms set in, then that entrenched system behavior changes quickly.

A mindset shift for sustainability is following Dr. King's advice and going from a focus on things to a focus on people; from consumers to citizens; from exploiting limited resources to nurturing unlimited resourcefulness; from chasing the mirage of unlimited material growth to finding unlimited nonphysical growth.

A shift is a young Bhutanese king prioritizing gross national happiness above gross domestic product. A shift is extending rights to future generations. It's caring about our children and grandchildren, but not at the expense of others. A shift is my parents more than offsetting their carbon footprints to leave an ecological inheritance, not just an economic one.

Mindset shifts are going from controlling Nature to learning from it; from destroying the Everglades' natural systems to amplifying them; from living off the land to living with the land; from humans as separate from Nature to humans as part of it; and from humans or apes to humans are apes.

A mindset shift is going from the environment as an accounting afterthought to recognizing that without the environment to sustain human life, economic accounting is pointless. From not-my-problem externalities to companies and consumers bearing the full social costs and benefits of their actions.

Mindset shifts are going from linear use and waste to cyclical closed loops (with T-shirt quilts); from extracting stock-limited oil resources to harnessing relentless wind renewables; from recycling as cure-all to recycling as last resort.

In our built environment, mindset shifts are moving from needless new buildings to elegant uses of what we already have; from electric bills as money-collector to electric bills as consumer education; from wondering what we can add to seeing what we can remove; and from

31. I reminded myself that mindsets are the greatest leverage point whenever I wondered why I was spending so much time writing this book.

jetties to dune grass. It's a mindset shift when we realize that roads should serve people (not automobiles); and it's a shift when we consider not only what these roads connect, but also what they divide.

A mindset shift is going from worrying about the unknowable to learning how to quickly regrow our tails should the need arise.

A mindset shift is moving from equality as the right approach to equality as the best approach; from hierarchy to cooperation; from competition as the enemy to competition as the fastest way to improve; and from thinking that men are the only suitable heads of state to realizing that this is a stupid (and statistically impossible) idea.

In our policies, mindset shifts are moving from complex rules to control behavior to the simple rules that let complex behavior thrive; from incremental rules to game-changing leverage points; and from arguing over the false dichotomy of private versus public to managing for the common good (as lobstermen already do).

As we create, mindset shifts take us from models for impossible clairvoyance to models for endless learning; from focusing on four-year election cycles to considering seven generations; from a narrow focus on what's easy to measure to a broad view of what we really care about; and from reducing unsustainability toward creating true sustainability.

Instead of chasing more sustainable hamburgers, we find more sustainable food. And instead of the perilous monoculture, we create the adaptable and resilient food system.

Mindset shifts for sustainability are seeing new connections: between climate change and human rights; between beauty and function; between science and spirituality[32] (and soccer); and between hundreds of stranded polar bears and hundreds of millions of displaced humans.

So before I let you go, please imagine a well-meaning Brazilian father whose daughter is born in 2005. The father is prejudiced with the prevailing mindset in Brazil. He believes that soccer is only for men, and so he plans to discourage his daughter from playing the Game.

But then the father hears about Marta winning World Player of the Year in 2006. He sees Marta lead Brazil to the World Cup final in 2007. She wins World Player of the Year that year too, and again in 2008 and 2009—and once more in 2010.

32. For me, spirituality is the nonmaterial things, in particular those moments when mind and body are harmoniously aligned. Or, as the Dutchman Cruyff put it, "I think clarity of spirit almost always comes with sport."

4

The father realizes that his precious daughter can play soccer after all. He realizes that gender equality is an opportunity and not just a requirement. Marta's example changes the father's mindset, and when that happens, soccer *can* be more than a game.

And the father's mindset has changed just in time, because now that his daughter is four years old, she's ready to learn how to juggle that tennis ball with her feet.

Mindset shifts are the greatest leverage points and why Marta's influence rivals that of Bertha and Dilma.

Review: The Endless Quest

A vision is our dream for how we want things to be in the future. Our visions guide us when we encounter unforeseen decisions and obstacles on our sustainability quests.

When they work as intended, rules can be dynamic disturbances that improve systems and move us toward our visions.

Leverage points are places where a small shift in one thing can produce big changes in everything. We can push leverage points in all the parts of a system: the elements, flows, stocks, purposes, and mindsets.

Mindsets are the values, ideas, and shared assumptions that are the foundations of society (and soccer).

Glossary

ADAPTABILITY The ability to perform multiple functions and to perform the same function multiple ways.

AUTONOMY The ability to act independently, free from external influence or control.

BIOMIMICRY Asking "how has Nature already solved this challenge?" and then applying what we find.

CARRYING CAPACITY The population that an environment can sustain indefinitely, given the available resources.

CLOSING LOOPS Eliminating waste and ending dependence on inputs from outside a system.

COMMON POOL RESOURCE Resources for which users get the benefits, but costs are shared by a larger group, regardless of use.

COUNTERFACTUAL ANALYSIS A systematic way of asking "what if" questions in order to consider alternative outcomes.

DIVERSITY The number and variety of relationships and approaches.

ELEGANCE Surprising simplicity on the other side of complexity.

ELEMENT The basic building blocks of a system.

EMERGENCE When entirely new system behaviors arise from simple organizing rules and resulting interactions.

ENVIRONMENTAL JUSTICE Equal protection from environmental hazards and equal access to the decision-making process that determines living conditions.

EXTERNALITY When the people getting the benefits are not the same people who are paying the costs, and vice versa.

FEEDBACK LOOP When a stock, actions dependent on the stock, and a flow or element to change the stock are connected. Balancing feedback loops slow down the rate of change, and reinforcing feedback loops speed it up.

FIVE WHYS Repeatedly asking "why" in order to uncover new perspectives and details.

FLOW Movements through which system elements interact.

INERTIA A resistance to change in the state of motion.

INPUT Flows to a system from its surroundings.

INTERDEPENDENCE How each part of a system gets its behavior from, and owes its existence to, its relationships to the other parts of the system.

LEVEL OF DETAIL The decision about how to accurately represent a system without wasting effort and without distracting from why we're interested in the system in the first place.

LEVERAGE POINT Places in a system where a small shift can produce big changes.

LIFE-CYCLE ANALYSIS A method to discover the value added by a product or service as well as the resulting sustainability impacts over a period of time.

MINDSET The values, ideas, and assumptions from which our systems arise.

OUTPUT Flows from a system to its surroundings.

PATH DEPENDENCE When present circumstances are limited by past decisions, even when past circumstances are no longer relevant.

POLYCENTRISM When multiple people or groups work together to manage a shared resource.

PURPOSE What the system is trying to accomplish, as indicated by its behavior.

REBOUND EFFECT When efficiency gains lead to unexpected new behaviors that negate some of the initial gains.

REDUCTIONISM Taking things apart (not always literally) and then studying the pieces in more detail.

REDUNDANCY Extra elements, flows, and resources that provide stability but delay the need to adapt.

RESILIENCE Minimizing damage from an unexpected event and then, after surviving the first impact, bouncing back.

RULES Dynamic disturbances that can improve systems when they work as intended.

SELF-ORGANIZATION When initial rules combined with interactions between elements (as opposed to top-down control) drives system behavior.

STOCK Built-up elements and flows in a system.

SUSTAINABILITY Meeting the needs of the present without compromising the ability of future generations to meet their own needs.

SYSTEM Connected elements and flows and the resulting behavior.

SYSTEM BOUNDARIES The system space, time, and level of detail.

SYSTEMATIC INEQUALITY Imbalances that are perpetuated by the system.

SYSTEMS-THINKING A shift in perspective from the parts to the whole, from objects to relationships, and from structures to processes.

THRESHOLD The point at which a system moves from one state to another.

TRANSPARENCY Access to the right information, in the right format, at the right times.

VISION Your dream for what the more sustainable future will look like.

Recommended Reading

Rather than try (and fail) to list every last reference for the wide range of topics covered, I've shared the best that I have found for many of the sections of this book. I hope this approach meets your desire to learn more and frees you to explore these topics on your own, without my prescription.

1. BACKGROUND

My Delayed Epiphany about Sustainability

For on-the-ground reporting of the link between climate change and human rights, read Christian Parenti's *Tropic of Chaos: Climate Change and the New Geography of Violence.*

Why Soccer? Sócrates Has Our Answer

If you're still not convinced of "why soccer" after this section, read *Soccer in Sun and Shadow* by Eduardo Galeano. Also read it if you are convinced.

The Black Panther and the Sacred Monster—Systems-Thinking for Sustainability

The holy text for systems thinking is *Thinking in Systems: A Primer* by Donella Meadows.

Why Me? An Autobiography in Less Than Twelve Hundred Words

Thankfully, there is no expanded autobiography of me, so allow me to recommend the book that has most shaped who I am: Mihaly Csikszentmihalyi's *Flow: The Psychology of Optimal Experience.* Read it only if you want to be happier and more productive.

2. PARTS

People and Players—System Elements

Elements are pretty self-explanatory, so no additional sustainability reference is needed here. But my favorite account of Brazil's soccer history is *Futebol: The Brazilian Way of Life* by Alex Bellos.

Floating Jabulanis and the River of Grass—Overlooked System Elements

Among the growing literature about measuring ecosystem services and other overlooked things that matter, you can't go wrong with *Natural Capitalism: Creating the Next Industrial Revolution* by Amory Lovins, Hunter Lovins, and Paul Hawken.

"Remember Istanbul" and the Cross Bronx Expressway—Physical Flows

For an overview of city-scale sustainability, read *Green Metropolis: Why Living Smaller, Living Closer, and Driving Less Are the Keys to Sustainability* by David Owen.

To learn more than any non-manager should know about soccer tactics, read Jonathan Wilson's *Inverting the Pyramid: The History of Soccer Tactics.*

Smoking Managers and Electric-Bill Peer Pressure—Information Flows

You could spend a year looking at all the academic research on how feedback shapes energy-use behavior. Save yourself the time because experts already did—and they published their summary in "Advanced Metering Initiatives and Residential Feedback Programs: A Meta-Review for Household Electricity-Saving Opportunities." For this readable synthesis, we can thank Karen Ehrhardt-Martinez, Kat Donnelly, Skip Laitner, and the American Council for an Energy-Efficient Economy.

Porto/Chelsea and a Russian "Oil"garch—Stocks

An agendaless take on energy policy in the United States is Michael Levi's *The Power Surge: Energy, Opportunity, and the Battle for America's Future.*

An illuminating look at the history that has given us oilgarch soccer owners is Jonathan Wilson's *Behind the Curtain: Football in Eastern Europe.*

Obsessive Shooting Practice and Population Bombs—Feedback Loops

You'll respect the power of feedback loops even more if you read *Social Physics: How Good Ideas Spread—The Lessons from a New Science* by Alex Pentland.

Falling Balls in Baltimore and Happy Bhutanese—Purposes

For more on measuring happiness, a good place to start is *The Happiness Manifesto* by Nic Marks.

3. BOUNDARIES

"El Loco" Higuita and the Nine-Dots Puzzle—Space Boundaries

To teach yourself, and others, techniques to think outside the box, read *Out of Our Minds: Learning to Be Creative* by Ken Robinson.

Glory-Days Sour Grapes and Seven Generations—Time Frames

Legal templates to ensure that future generations have the same opportunities we do are in the report "Models for Protecting the Environment for Future Generations" by the Science and Environmental Health Network and the International Human Rights Clinic at Harvard Law School.

A Bent-Legged Angel and Sustainable Hamburgers—Level of Detail

Perhaps the sustainable hamburger is not pure greenwashing? Expand your perspective with *Defending Beef: The Case for Sustainable Meat Production* by Nicolette Hahn Niman.

Icelandic Volcanoes and the Best Team Ever—Inputs and Outputs

If you still deny climate change, please read *The Weather Makers: How Man Is Changing the Climate and What It Means for Life on Earth* by Tim Flannery.

And if you still doubt Barcelona's dominance, read *Barca: The Making of the Greatest Team in the World* by Graham Hunter.

Greece vs. Europe and Ehrlich vs. Simon—Insight, Not Clairvoyance

For more on what models can and cannot tell us, I recommend Nate Silver's *The Signal and the Noise: Why So Many Predictions Fail—But Some Don't.*

4. BEHAVIORS

The Worst Game Ever and Martin Luther King Jr. as an Environmentalist—Interdependence

Environmental justice is a central theme in Pope Francis's *Encyclical on Climate Change and Inequality: On Care for Our Common Home.*

Pickup Games and Hungry Ants—Self-Organization

For more on self-organization, read *Linked: The New Science of Networks* by Albert-László Barabási and Jennifer Frangos.

For more on pickup soccer, read *Finding the Game: Three Years, Twenty-Five Countries, and the Search for Pickup Soccer* by Gwendolyn Oxenham, who also produced *Pelada,* a beautiful documentary on the same topic.

Mayan Ball Games and Chimpanzees—Emergence

Read about all kinds of emergent phenomena in *The Systems View of Life: A Unifying Vision* by Fritjof Capra and Pier Luigi Luisi.

Unfair Goals and Lewis's Lizard—Resilience

If you're looking for non-lizard examples of resilience, read *Resilience: Why Things Bounce Back* by Andrew Zolli and Ann Marie Healy.

Zidane and Disappearing Ice—Threshold Crossing

Stay current on projected climate-change implications with the comprehensive and readable summaries produced by the nonpartisan Intergovernmental Panel on Climate Change (IPCC). Google "IPCC Assessment Reports."

5. EVALUATING

Lampard and Me, Highways and Railroads—Path Dependence

To have the future-shaping power of path dependence seared into your brain, read *The Other Wes Moore: One Name, Two Fates* by Wes Moore.

Panenka's Gift and New Jersey Dune Grass—Inertia

Want to know more about why you like your favorite vacation spot? Read Richard Louv's *The Nature Principle: Reconnecting with Life in a Virtual Age.*

Argentine Defenders and Unsuicidal Lemmings—Carrying Capacities

For details about carrying capacities (but not why Zanetti was left off of Argentina's 2010 World Cup team), read *Limits to Growth: The 30-Year Update* by Donella Meadows, Jorgen Randers, and Dennis Meadows.

My Missed Penalty and a Stern Review—Counterfactuals

To imagine how things could be better where you live, read *Clean Break: The Story of Germany's Energy Transformation and What Americans Can Learn from It* by Osha Gray Davidson. To imagine how things could be worse, read *Bird on Fire: Lessons from the World's Least Sustainable City* by Andrew Ross.

Barbosa, Bigode, and the Choice to Eat Dirt—The Five Whys

An indispensable synthesis of decision-making tools, and the science behind them, is *Decisive: How to Make Better Choices in Life and Work* by Chip Heath and Dan Heath. I also recommend their books on communication *(Made to Stick)* and making change *(Switch).*

Footprints of the World Cup—Life-Cycle Assessment

My favorite life-cycle assessment book is *How Bad Are Bananas: The Carbon Footprint of Everything* by Mike Berners-Lee. Because of this book, I use disposable diapers for my son and I rarely hand wash dishes. If you're a banana-loving environmentalist, don't worry—bananas are pretty good compared to most foods. To learn more about life-cycle assessment without buying a book (or being entertained), start with the U.S. Environmental Protection Agency's "LCA 101: Life Cycle Assessment: Principles and Practice."

6. CREATING

False-Brooding Runs and Wind Turbines—Biomimicry

No one has done more to popularize biomimicry than Janine Benyus, in part through her book *Biomimicry: Innovation Inspired by Nature.*

Soccer-Shirt Quilts and the Recycling Distraction—Closing Loops

To get ideas for closing loops on a scale larger than t-shirts, try Austin Troy's *The Very Hungry City: Urban Energy Efficiency and the Economic Fate of Cities.*

The Goalkeeper Pick Trick and Irish Lumpers—Adaptability

Jared Diamond's *Collapse: How Societies Choose to Fail or Succeed* will make you wonder how adaptable we are.

"Los Galácticos" and New England Lobstermen—Polycentrism

Elinor Ostrom's book *Governing the Commons: The Evolution of Institutions for Collective Action* is a firsthand account of her research.

The Fan Who Scored for West Ham, and Divestment—Transparency

Articles rather than books are best here. For more on fossil fuel divestment, Google "Bill McKibben divestment article." He regularly publishes his insights relevant to the ongoing campaign.

"The Day Harry Redknapp Brought a Fan on to Play for West Ham" is Jeff Maysh's perfectly titled retelling of Steve Davies's goal for his favorite club. The article originally appeared in *Howler* magazine, and you can find it online with a Google search.

"Take the Piss" and "Let Them Eat Cake"—Fixing Inequality

Branko Milanovic weaves an entertaining history of inequality in *The Haves and the Have-Nots: A Brief and Idiosyncratic History of Global Inequality.*

Stefan Szymanski delves into the history of inequality in soccer in *Money and Soccer: A Soccernomics Guide.*

Breakaways, Passbacks, and My Repurposed Office—Elegance

An inspiration for my office change was *Make Space: How to Set the Stage for Creative Collaboration* by Scott Doorley and Scott Witthoft.

7. THE ENDLESS QUEST

For an in-depth and comprehensive vision for sustainability, read *Carbon Zero: Imagining Cities That Can Save the Planet* by the self-proclaimed futurist Alex Steffen.

The best account of Dutch soccer, Total Football, and Johan Cruyff's role in it is *Brilliant Orange: The Neurotic Genius of Dutch Soccer* by David Winner.

Offsides and Refrigerators—Rules

A surprisingly entertaining read about energy efficiency is Stan Cox's *Losing Our Cool: Uncomfortable Truths about Our Air-Conditioned World (and Finding New Ways to Get Through the Summer).*

Bertha, Dilma, and Marta—Leverage Points

As with so many other ideas, Donella Meadows artfully explains "leverage points" in the systems context. You can find her take by Googling "Donella Meadows leverage points."

The Greatest Leverage Point and Marta Continued—Mindsets

This Changes Everything: Capitalism vs. the Climate by Naomi Klein has a misleading title but makes a convincing case for the power (and necessity) of changed mindsets.

Index

Abramovich, Roman, 35–36, 92–93
AC Milan, 8n17, 26–27
accounting, 67n4, 162. *See also* evaluation
adaptability, 79, 86, 128–131; definitions and key ideas, 129, 131, 146, 165; soccer analogies, 128–129, 131, 150, 161
Adidas, 59; the Jabulani ball, 20–22, 24–25, 124n11
African soccer teams and players, 7–8, 47–48. *See also specific countries and players*
agriculture, 110–112, 120n2, 163; the Irish potato famine, 129–130
air pollution, 71
air travel, 113, 116
Ajax Amsterdam, 149
alternative outcomes. *See* comparisons; counterfactual analysis
American football, 19n10, 30, 79
American soccer. *See* U.S.
Ancelotti, Carlo, 34n40
ant behavior, 75–76
anthropocentrism, 5n14
anti-apartheid divestment campaign, 136–137
Antoptima, 76–77n26
ape evolution, 79–80
Arctic ice, 38n44, 85–86
Argentine soccer, 18n8, 72, 99–102, 150n6
AskNature, 124n11
assessments. *See* evaluation
association football, 78, 79
assumptions, 107. *See also* mindsets
asthma rates, 71
autonomy, 165. *See also* polycentrism

balancing feedback loops, 17, 37, 38, 142, 165
ball games, 77–78
Ballon d'Or, 20n12, 149
balls: the Jabulani, 20–22, 24–25, 124n11
Barbosa, Moacir, 108–109, 111
Barcelona. *See* FC Barcelona
barrier islands, 97
basketball, 30–31, 145n51
Bayer Leverkusen, 132n30
Bayern Munich, 100
beach erosion, 97–98
Bearzot, Enzo, 30
Beckham, David, 34n39, 131, 132
beef, 56–58, 114, 115
behaviors, 66–88; elements as influences, 19; emergence, 68, 77–80, 88, 165; flows as influences, 27, 29; interdependence, 4–5, 68–74, 88, 166; overview, 66–69; purposes as influences, 41; resilience, 68, 80–83, 88; review, 88; self-organization, 67–68, 74–77, 88, 166; thresholds and threshold crossing, 68, 83–87, 88; time frames and, 51; uncertainty of future behaviors, 46, 61–65

Benfica, 7, 8, 16–17, 45, 46, 67–69, 120–122, 161
Benítez, Rafa, 31–32
Bent-legged Angel (Manuel Francisco dos Santos), 53–55, 140
Berbatov, Dimitar, 97n15
Betancur, Belisario, 72
betting and odds, 62–63
Bhutan, 41–42, 151
Bielsa, Marcelo, 132n29
Bigode (João Ferreira), 109, 111
biodiversity, 130n25
biomimicry, 120, 122–125; definitions and key ideas, 123, 125, 146, 165
Black Panther (Eusébio da Silva Ferreira). *See* Eusébio
Blatter, Sepp, 157, 161
Bolivian constitution, 52
Bolivian soccer, 59n23, 73n22
bonobos, 79
boundaries, 44–65; definitions and key ideas, 44–46, 65, 166; future uncertainty and, 46, 62, 65; inputs/outputs and, 61, 65; level of detail, 45, 46, 53–58, 65, 166; the nine-dots puzzle, 48–50; overview, 44–46; review, 65; spatial boundaries, 44–45, 46–50, 65; time frames, 45–46, 50–53, 65. *See also* uncertainty
Bradley, Bob, 12n25
Brazil: childhood dirt eating in, 110–112; gender equality and women's rights in, 156–161, 163–164; Roussaeff and her election to the presidency, 158–159; as 2014 World Cup host, 112–113; Vargas's dictatorship and suicide, 108n32, 158–159
Brazilian Federation for the Advancement of Women, 158
Brazilian soccer, 18–19, 156; Brazil-Argentina rivalry, 54n13; Marta Vieira da Silva, 156–157; 1950 World Cup, 107–110, 111, 161; 1958 World Cup, 54–55, 73, 111; 1966 World Cup, 91; 1982 World Cup, 4n12; 1994 World Cup, 72n16; psychological testing of players, 54–55; World Cup teams, 54–55. *See also specific clubs and players*
British soccer. *See* English soccer; *specific clubs and players*
Brito, Maurício Vieira de, 69
Bronckhorst, Giovanni van, 21n14
Brundtland Commission, 66–67n3
Buffalo Stallions, 12n24
Cameroon, 47–48, 145
Camus, Albert, 4
Capello, Fabio, 21
carbon emissions and levels. *See* CO_2
carbon footprint assessments, 116–117. *See also* life-cycle analysis
carbon neutrality, Costa Rica's vision for, 151, 153
carbon sequestration, 23, 111n38
carbon taxes, 29
Carbon Tracker Initiative, 137n38
Carlin, George, 67n2
carrying capacity, 90, 99–104; definitions and key ideas, 102, 104, 118, 165
cars and roads, 28–29, 93–94, 163
Casey, Susan, 125n14
Catt, Carrie Chapman, 158
causes, uncovering. *See* five whys
Champions League. *See* European Cup/Champions League
change, rate of. *See* rate of change
change, resistance to. *See* inertia
Chapman, Lee, 135, 136, 139
Chelsea FC, 33n38, 34–37, 92–93
Chile, 69–70, 72, 73, 74
chimpanzees, 79
Chinese family size rules, 39
civil rights movement, 70
clairvoyance. *See* uncertainty
Clemson, South Carolina, 27–28
Clemson University: author's office improvements, 144; fossil-fuels divestment and, 137, 139
climate change, 6, 147; as a security threat, 2; carbon sequestration and, 23n19; climate-change thresholds, 85–86; climate justice, 147n1; climate refugees, 1–2; economic costs of, 86, 106, 107; feedback loops in, 38n44, 86; fisheries impacts, 133; unequal impacts of, 71; unpredictable impacts of, 86n41, 120n2; war and, 103n25
closing loops, 120, 125–128; definitions and key ideas, 126, 128, 146, 165
CO_2: carbon sequestration, 23, 111n38; current atmospheric CO_2 levels, 138n40; emissions from different electricity sources compared, 123n9; ocean acidification and, 85n39; potential impact of burning fossil-fuel reserves, 137. *See also* carbon *entries*; climate change
coal, 123n9
collaboration, 121. *See also* polycentrism

college endowments, 137
Colombia: fisheries management in, 134n36; 1990 World Cup, 47–48, 145; relinquishment of 1986 World Cup hosting duties, 72
Coluna, Mário "the Sacred Monster," 8, 15–17, 45, 67–69; and Eusébio, 16, 17, 46, 68, 161; 1961 European Cup final, 67–68, 69; 1962 European Cup final, 120–122; 1966 World Cup, 89–90, 91
commodity prices: the Ehrlich-Simon bet, 63, 65
common pool resources, 133–135, 165
Common Wealth: Economics for a Crowded Planet (Sachs), 3n4
comparisons: of alternative outcomes, 106, 107, 153; sustainability choices and, 32–33, 57
competition, 72–73
consumption levels, Earth's carrying capacity and, 103
CoolClimate Network carbon footprint calculator, 117n48
cooperation, 163
Corinthians, 4n12, 73. *See also* Sócrates
corner kicks, 122–123
corporate rights and decision making, 53
Costa Rica, 151, 153
counterfactual analysis, 91, 104–107; definitions and key ideas, 105, 107, 118, 165
creating systems, 119–146; adaptability, 79, 86, 128–131, 146, 165; biomimicry, 120, 122–125, 146, 165; closing loops, 120, 125–128, 146, 165; correcting inequality, 121, 139–142, 146, 166; elegance, 79, 121–122, 142–146, 165; overview, 119–122; polycentrism, 71, 120–121, 131–135, 146, 166; review, 146; transparency, 121, 135–139, 146, 166. *See also* sustainability quests
Cross Bronx Expressway, 28–29
Crouch, Peter, 77n27
Cruyff, Johan, 30, 120n3, 149–151, 152–153; quotes from, 72n19, 145, 150, 163n32
Czechoslovakian soccer: Panenka and the Panenka technique, 84–85, 95–96, 98, 99, 161

DaMatta, Roberto, 108
Dassler, Adi, 59n22
Davies, Steve, 135–136, 153
decison-making ability and processes, 58, 71
Del Bosque, Vicente, 131–132, 135, 152, 161
Dempsey, Clint, 21n14
Desailly, Marcel, 7n15
detail. *See* level of detail
Di Matteo, Roberto, 34n40
dirt eating, 110–112
discrimination, 140. *See also* inequality
diversity, 82–83, 111–112, 129–130, 165
divestment campaigns, 136–139
Dos Santos, Manuel Francisco "Garrincha," 53–55, 140
Downs, Anthony, 29n30
drought, 120n2
dummy runs, 122–123
Dutch soccer teams, 120–122, 150, 152–153
Dutch West India Company, 111n39
dynamism, 16, 80
Dynamo Kiev, 148

Earth, carrying capacity of, 103
economic inequality, 140–142
economic sustainability/health, 67
ecosystem services, 23–24, 38
Ecuadoran constitution, 52
Ehrlich, Paul, 63, 65
Einstein, Albert, 9
El Loco (René Higuita), 46–48, 49, 145, 161
El Salvador, soccer in, 4n8
electricity: supply technologies compared, 123n9, 124n10; wind energy, 123–125. *See also* energy *entries*
electricity bill comparisons, 32–33
elegance, 79, 121–122, 142–146; definitions and key ideas, 144, 146, 165
elements, 18–25; definitions and key ideas, 15, 19, 22, 43, 165; overlooked elements, 20–25; soccer analogies, 15–16, 18–22, 24–25, 50
emergence, 68, 77–80; definitions and key ideas, 79, 80, 88, 165
the endless quest. *See* sustainability quests
energy costs: Brazilian energy tax, 160n30
energy efficiency standards, 154, 155
energy-efficient stadium designs, 112–113, 114
energy policy, sustainable, 37
energy resources: CO_2 emissions compared, 123n9; renewable vs. nonrenewable, 35, 37; wind energy, 123–125
energy use and demand, 32–33, 94, 113, 124n10

English soccer, 27n24, 59n24; the Cambridge Rules, 78–79; Frank Lampard, 92–93; 1962 World Cup, 90; 1966 World Cup, 89–90; 1986 World Cup, 100; 2006 World Cup, 77n27. *See also specific clubs and players*
environmental justice, 23–24, 70–71, 165; climate justice and the plight of climate refugees, 1–2, 147n1
Epictetus, 81n30
episkyros, 4n11, 78
equality, 163. *See also* inequality
Escobar, Pablo, 46
Eton Rules, 78
Euro 2004, 61–62, 63n31, 64
European Cup/Champions League, 25; FC Barcelona in, 60–61; 1961, 67–68, 68–69; 1962, 121; 1963, 68; 1976, 95–96; 1992, 150–151; Real Madrid in, 132; 2000, 84n37; 2002, 132n30; 2004, 34; 2005 final (Istanbul), 26, 31–32, 82, 106n29; 2009, 152; 2010, 60–61, 100, 105n21
Eusébio (Eusébio da Silva Ferreira "the Black Panther"), 7, 8, 90, 106n29; and Coluna, 16, 17, 46, 68, 161; 1962 European Cup, 121; 1966 World Cup, 91; U.S. indoor league play, 12n24, 40n46
evaluation, 89–118; carrying capacity, 90, 99–104, 118, 165; counterfactual analysis, 91, 104–107, 118, 165; the five whys, 91, 107–112, 118, 166; inertia, 89, 95–99, 118, 166; life-cycle analysis, 56–57, 91, 112–117, 118, 166; overview, 89–92; path dependence, 89–90, 92–95, 118, 166; review, 118; triple bottom line accounting, 67n4
Everglades development, 23–24, 25, 35n41
evolution, 79–80
externalities, 28–29, 165
extreme weather, 71, 86. *See also* flooding
Eyjafjallajökull volcano, 60–61, 105n27

false brooding, 123
Faria, Romário de Souza, 72
farming. *See* agriculture
FC Barcelona, 33nn37,38, 60–61, 100, 105n27, 161; Cruyff and his youth development system, 150–151, 152–153
FC Bayern Munich, 100
FC Porto, 34, 36, 37
Fédération Internationale de Football Association. *See* FIFA
feedback loops, 37–39; balancing vs. reinforcing, 17, 37, 39, 43, 165; climate change and, 38n44, 86; definitions and key ideas, 16–17, 37, 39, 43, 165; as inequality corrections, 142
Ferreira, Eusébio da Silva. *See* Eusébio
Ferreira, João "Bigode," 109, 111
FIFA (Fédération Internationale de Football Association), 22, 30, 69, 74, 113, 157
Figo, Luís, 131, 132
Filho, Mário, 73n21
financial divestment campaigns, 136–139
Finnan, Steve, 26
fish schooling behavior, 124
fisheries management, 133–135
five whys, 91, 107–112; definitions and key ideas, 110, 112, 118, 166
flooding, 45, 46, 86, 97, 120n2; in Maputo, 46, 67, 119, 120, 121
Florida Everglades, 23–24, 25, 35n41
flow-limited resources, 35
flows, 25–33, 43; definitions and key ideas, 16, 27, 32, 33, 43, 166; as influence on stocks, 159–160; of information, 30–33, 43, 75, 159; in life-cycle analysis, 115; physical, 25–29, 43; restricting with rules, 39; soccer analogies, 16, 25–27, 30–32, 37
Fogel, Robert, 105
food systems, 55–58, 86, 163. *See also* agriculturepoverty and nutrition, 110–112
football, 4n10. *See also* American football; soccer
Forlán, Diego, 21n14
fossil fuels, 123n9, 124n10; divestment campaigns, 136–139. *See also* CO2; energy *entries*
fouls, 84–85, 86–87. *See also* penalty kicks
"frantic city," 81n31
French Revolution, 140–141
French soccer teams, 7n15, 84–85, 86–87. *See also specific players*
functional units, for life-cycle analyses, 115
future generations, considering (long time frames), 51–53, 162

Galácticos (Real Madrid), 131–132, 135
Galeano, Eduardo, 21, 77, 155n21
Garcé, Ariel, 101
Garrincha (Manuel Francisco dos Santos), 53–55, 140
Gaviotas: A Village to Reinvent the World (Weisman), 134n36

gender equality, 156–161, 163–164
German soccer, 5n13, 100n19; and the Jabulani, 24n21; 1976 European Championship, 95–96; World Cup teams, 59, 100, 150
Gerrard, Steven, 26, 27
Ghana national team, 47n7
Ghiggia, Alcides, 109
goalkeeper pick trick, 128–129, 141
goalkeepers, 48, 96, 129; pass-back rule change, 144–145. *See also* penalty kicks; *specific players*
goals and goal scoring, 31, 105, 143–144; scoring and rule changes, 144–145, 154, 155. *See also* penalty kicks
Gonçalves, Delma (Pretinha), 157
Grant, Avram, 34n40
Greece, ancient, 4, 78
Greek soccer team, in Euro 2004, 61–62, 63n31, 64
Green, Robert, 21n14
Greenland ice sheet, 85n40
greenwashing, 113–114
growth, 67
Gutiérrez, Jonás, 101

Hamann, Dietmar, 26
hamburgers, 56–58, 114, 115
happiness, valuing and measuring, 41–43, 151, 162
Happy Planet Index, 41–42, 151
Harkes, John, 45n1
Harrisburg Heat, 12n24, 18, 40n45, 128–129
Harvard University endowment, 137
Hayes, Chris, 147n1
Hejduk, Frankie, 12n25
Henry, Thierry, 97n15
Hiddink, Guus, 34n40
Higuita, René “El Loco”, 46–48, 49, 145, 161
hockey, 145n51
home-field advantage, 90nn2,4
Honda, Keisuke, 21n14
Honduras, soccer in, 4n8
housing planning, 27–28
human evolution, 79–80
human health, 67; poverty and nutrition, 110–112
human impacts: carbon footprint measurements, 116–117; impacts in life-cycle analysses, 114–115, 115–116; measuring with IPAT, 103
human population levels, 38, 39, 103. *See also* carrying capacity
human rights: for future generations, 52–53, 162. *See also* environmental justice
human waste management, 46, 122
humpback whale, 124
Hungary, 52; Hungarian soccer teams, 51n9, 59
Hunt, Lamar, 35n42
Hurricane Katrina, 71
Hurricane Sandy, 86

ice sheets, climate change and, 38n44, 85–86
Iceland: Eyjafjallajökull volcano, 60–61, 105n21
IFAB (International Football Association Board), 145n50
indigenous ball games, 77, 78
indoor soccer, 40, 129
inequality, 121, 139–142; definitions and key ideas, 140, 141–142, 146, 166; gender inequality, 156–161, 163–164
inertia, 89, 95–99; definitions and key ideas, 96, 99, 118, 166
information. *See* information flows; transparency
information flows, 30–33, 43, 75, 159; definitions and key ideas, 32, 33, 43. *See also* flows
Iniesta, Andrés, 152
inputs and outputs, 44, 59–61; definitions and key ideas, 61, 65, 166. *See also* closing loops
Inter Milan, 61, 99–100, 161
interdependencies, 4–5, 68–74; competition as, 72–73; definitions and key ideas, 70, 74, 88, 166; King on, 70, 73
Intergovernmental Panel on Climate Change (2014), 123n9
International Football Association Board (IFAB), 145n50
IPAT equation, 103
IPCC (2014), 123n9
Irish potato famine, 129–130
iron deficiency, 110–112
Iroquois confederacy, 52
Istanbul Champions League final (2005), 26, 31–32, 82, 106n29
Italian soccer, 27n24; 1934 World Cup, 72; 1938 World Cup, 110n37; 2006 World Cup, 84–85, 86–87. *See also specific clubs and players*

Jabulani ball, 20–22, 24–25, 124n11
Jacobs, Jane, 29n32

Japanese constitution, 52
Jennings, Andrew, 22n17
jetties, 97–98

Kewell, Harry, 26
King, Martin Luther, Jr., 70, 71, 73, 74, 162
Kuper, Simon, 108n33

La Volpe, Ricardo, 30
laminar flow, 21–22n15
Lampard, Frank, 92–93
LCA. *See* life-cycle analysis
lemming behavior, 102
"Letter from Birmingham Jail" (King), 70
level of detail, 45, 46, 53–58; definitions and key ideas, 55, 58, 65, 166
leverage points, 148–149, 156–161, 162; definitions and key ideas, 158, 161, 164, 166
Lewis's lizard, 82
life-cycle analysis, 56–57, 112–117; definitions and key ideas, 114, 117, 118, 166; electricity-supply technologies compared, 123n9
Linekar, Gary, 97n15, 100n19
Liverpool FC, 26–27, 31–32, 82, 106n29
Lobanovskyi, Valeriy, 148
lobster fisheries, 133–135
Loco (René Higuita), 46–48, 49, 145, 161
Los Galácticos (Real Madrid), 131–132, 135
Louis XVI, 140–141
Luisi, Paulina, 158
Lutz, Bertha, 158, 159, 160, 161

McAvennie, Frank, 92
McDonald's hamburgers, 56–58, 114, 115
McKibben, Bill, 137
Maier, Sepp, 96
Major League Soccer (MLS), 142–144, 145
Mandela, Nelson, 137, 147–148
Maputo, Mozambique, 15, 16, 17, 66; clean-water access and management, 45–46, 119, 120, 121, 122, 148; flooding in, 46, 67, 119, 120, 121; hierarchy of needs in, 67
Maradona, Diego, 100–101
Marie Antoinette, 140–141
Martí, José, 129–130
Mayan ball game, 77
Mayan greetings, 73n24
Meadows, Donella, 158n24
measurement, 6. *See also* evaluation
meat production and sustainability, 55–58
Menotti, César Luis, 30
Messi, Lionel, 20–21, 152
microorganisms, 5n14
migrants and refugees, 1–2, 102–103, 129, 148–149
Milla, Roger, 47–48, 145
mindsets and mindset shifts, 149, 161–164; definitions and key ideas, 162, 164, 166
mineral deficiencies, 110–112
MLS (Major League Soccer), 142–144, 145
money: happiness and, 42; high-priced soccer players, 33–34, 35–36
monocultures, 129–130, 163
Moses, Robert, 29n31
Mourinho, José, 33–34, 150n10
Mozambique, 15n1, 16, 16n2, 44, 66n1; independence and ensuing civil war, 16, 67, 90, 91, 148–149; soccer players from, 7–8, 15–17. *See also* Coluna, Mário; Eusébio; Maputo
Mussolini, Benito, 72
Mustache (João Ferreira), 109, 111
mutual funds, socially responsible, 138, 139

Napoli soccer club, 101
Nascimento, Edson Arantes do. *See* Pelé
Native American ball games, 77, 78
natural gas, 123n9
natural systems, 8–9; ecosystem services, 23–24, 38. *See also* biomimicry
needs and needs hierarchies, 66, 67
The Negro in Brazilian Football (Filho), 73n21
Netherlands: Dutch soccer teams, 120–122, 150, 152–153. *See also* Cruyff, Johan
"The New Abolitionism" (Hayes), 147n1
New England lobster fisheries, 133–135
New Orleans: Hurricane Katrina's unequal impacts, 71
New York City: the Cross Bronx Expressway, 28–29; economic costs of Hurricane Sandy, 86
Neymar, 98n18
nine-dots puzzle, 48–50
nonrenewable resources, 35. *See also* fossil fuels
North American Soccer League offsides rule, 154n16
North Korean soccer team, 89, 106n29
nutmegging, 12n25

ocean acidification, 85n39
Ocean City, New Jersey, 97–98

offsides rule, 143, 153–154, 155
olé chants, 54
Oliveira, Sócrates Brasileiro Sampaio de Souza Vieira de, 4
Ostrom, Elinor, 133–135
Otamendi, Nicholás, 101
outputs. *See* inputs and outputs
overlooked system elements, 20–25
Oxford City FC, 135–136

Palermo, Martín, 101
Pan American Conference of Women (1922), 158
Panenka, Antonín, and the Panenka technique, 84–85, 95–96, 98, 99, 161
parts. *See* system parts; *specific parts*
pass-back rule, 144–145
path dependence, 89–90, 92–95; definitions and key ideas, 93, 95, 118, 166
Pato, Alexandre, 98n18
Pelé (Edson Arantes do Nascimento), 4n9, 53, 54, 55, 72, 96n14, 100n20, 101n22
penalty kicks, 26nn24,26, 27; author's missed kick, 104–105; 1962 European Cup final, 120–121; the Panenka technique, 84–85, 95–96, 98, 99, 161; related rule changes, 142–144, 145
Pérez, Florentino, 132
Persie, Robin van, 98n18
Petras, Doug, 128–129
Peucelle, Carlos, 150n6
Philadelphia, 19nn9,10
physical flows, 25–29, 43. *See also* flows
Picasso, Pablo, 146
pickup soccer games, 74–75, 140, 141
Pinochet, Augusto, 69, 70, 74
Pirlo, Andrea, 21–22n15, 98n18
Pittsburgh nature rights ordinance, 52
Pittsburgh Riverhounds, 132n30
plants, 5n14
Platini, Michel, 101n22
polar bears, 85
political access and participation, 71. *See also* voting rights
politics, soccer and, 69–70, 72
"polluter pays" principle, 29
polycentrism, 71, 120–121, 131–135; definitions and key ideas, 133, 135, 146, 166
population levels (human), 38, 39, 103. *See also* carrying capacity
Porto soccer club, 34, 36, 37
Portugal, Mozambique's independence from, 16, 148
Portuguese soccer: Euro 2004, 62; 1962 World Cup, 90; 1966 World Cup, 7, 8, 46, 89–90, 91, 106n29. *See also specific clubs and players*
potato famine, 129–130
poverty: food access/nutrition and, 110–112. *See also* environmental justice
Pretinha (Delma Gonçalves), 157
primate evolution, 79–80
Prohibition, 155
psychological testing, of soccer players, 54–55
public transportation, 93–94
purposes, 40–43, 44, 161; definitions and key ideas, 17, 41, 43, 166
Puskás, Ferenc, 59n22

Qatar, as 2022 World Cup host, 35n43, 90n5, 113, 151n13
Quadros, Jânio, 72

racial differences, genetically determined, 19n11
racism and racial equality: soccer and, 73, 109n36. *See also* civil rights movement
rate of change: feedback loops and, 17, 37, 39, 43; thresholds as sudden changes, 86, 87, 88
Real Madrid, 33n38, 84n38, 131–132, 135
rebound effects, 155, 166
recycling, 126–127
Redknapp, Harry, 92n6, 135–136
reducing waste, vs. reusing or recycling, 127
reductionism, 6–8, 166
redundancy, 120, 130, 166. *See also* diversity
refrigerator energy efficiency standards, 154–155
refugees and migrants, 1–2, 102–103, 129, 148–149
reinforcing feedback loops, 17, 38, 43, 79, 160–161, 165
relativity theory, 9
renewable energy resources, 35
resilience, 68, 79, 80–83, 163; definitions and key ideas, 81, 83, 88, 166
resources: polycentric management of common pool resources, 133–135; resource scarcity and conflicts, 38, 39, 63; single-product economies, 130. *See also* carrying capacity
reuse, 126, 127
Rio de Janeiro: the 1950 World Cup final, 107–110, 111
River Plate soccer club, 150n6

roads and traffic, 28–29, 93–94, 163
Robinho (Robson de Souza), 21
Ronaldo, Cristiano, 21–22n15, 131
Roussaeff, Dilma, 158–161
rugby, 78, 79
rules, 76, 77, 148, 153–156; definitions and key ideas, 154, 156, 164, 166. *See also* soccer rules
Russia: Abramovich as Chelsea FC owner, 35–36; as 2018 World Cup host, 35n43, 113. *See also* Soviet soccer

Sacchi, Arrigo, 150n10
Sachs, Jeffrey, 3n4
Sacred Monster. *See* Coluna, Mário
same-sex marriage, 160n30
sand dunes, 98
Santana, José Roberto Gomes (Ze), 18–20
Santos soccer club, 4n9
Scolari, Luiz, 34n40
scope (of a life-cycle analysis), 114–115
scorpion kick, 47, 48
sea-level rise, 86, 119
self-organization, 67–68, 74–77, 160; definitions and key ideas, 75, 77, 88, 166
Senegal national taeam, 47n7
sex, 64
Silva, Marta Vieira da, 156–157, 161, 163–164
Simões, António, 45
Simon, Julian, 63, 65
Smart Growth America, 93n8
smoking, on soccer sidelines, 30
soccer, 3–5; Dutch "Total Football," 150, 161; pickup games, 74–75, 140, 141; social-justice impacts of, 73. *See also* World Cup soccer; *specific clubs, countries, and players*
Soccer America, 92, 93
soccer balls: the Jabulani, 20–22, 24–25, 124n11
soccer fans, 22, 26, 54, 108; Davies's substitution and goal for West Ham, 135–136, 153; female fans, 157; as influence on play and players, 59, 82, 90n4, 109; U.S. MLS fans, 143, 145n51
soccer field conditions, 59
Soccer in Sun and Shadow (Galeano), 21. *See also* Galeano, Eduardo
soccer moms, 125–126
soccer players: adaptability to different roles, 150, 161; autonomy on the field, 131–132, 135; female players, 156–157, 161, 163–164; height and heading ability, 122; high-priced players, 33–34, 35–36; jersey advertising, 143n48; player development, 33–37, 111, 150, 152–153; player transfers, 92–93; psychological testing of, 54–55; red-card ejections, 84–85, 87; "taking the piss," 139–140; team captains' influence, 45. *See also* goalkeepers; *specific players*
soccer referees, 51, 59, 155
soccer rules, 78–79, 145n50; offsides rule, 143, 153–154, 155; the pass-back rule change, 144–145; tie-breaking shootouts or overtime, 143
soccer-shirt quilts, 126
soccer team management, 30–32; Cruyff's vision, 150–153; dividing players into teams, 128–129, 141; polycentrism, 131–132, 135; Redknapp's substitution with a fan, 135–136; World Cup rosters and, 100–101. *See also specific clubs, managers and coaches*
Soccer War, 4n8
Soccernomics (Kuper and Szymanski), 108n33
social inequality, 140–142. *See also* inequality
social sustainability, 67
Sócrates, 4
South Africa: apartheid-era divestment campaigns, 136–137; 2010 World Cup, 20–22, 113, 135
South Korea, in 1994 World Cup, 73n22
Souza, Robson de (Robinho), 21
Soviet soccer: Lobanovskyi, 148; 1974 World Cup team, 69–70
Spanish soccer: national teams and World Cup play, 83, 135, 152. *See also specific clubs and players*
spatial boundaries, 44–45, 46–50; key ideas, 48, 50, 65
spirituality, 163
stability, 38. *See also* balancing feedback loops; rate of change
stadium design and energy consumption, 112–113, 114
Stern, Nicholas, 106, 107
stock-limited resources, 35, 38
stocks, 33–37; definitions and key ideas, 16, 37, 43, 166; feedback loops as influences on, 37–39, 43; leverage points as influences on, 159–160
sugar production, in northeastern Brazil, 111

suicide: soccer-induced, 107, 108; Vargas's suicide, 108n32, 159n28
sustainability: definitions of, 5, 66–67, 166; long time frames and, 51
sustainability measures. *See* evaluation
sustainability quests, 147–164; leverage points, 148–149, 156–161, 164, 166; mindsets and mindset shifts, 149, 161–164, 166; overview, 147–149; review, 164; rules, 148, 153–156, 164, 166; vision, 148, 149–153, 164, 166. *See also* creating systems
Swindoll, Charles, 81n30
Syracuse Blitz, 92
Syrian migrants, 102–103
system parts, 15–43; elements, 15–16, 18–25, 165; information flows, 30–33, 43, 75, 159; overview, 15–17; physical flows, 25–29; purposes, 17, 40–43, 44, 161, 166; review, 43. *See also* feedback loops; stocks
systematic inequality. *See* inequality
systems, defined, 44, 166
systems-thinking, 5, 6–9; definitions and key ideas, 7*fig.*, 166
Szymanski, Stefan, 108n33

"taking the piss," 139–140
technology: and carrying capacity, 103
Terry, John, 45n1
350.org, 138
thresholds and threshold crossing, 68, 83–87, 160; definitions and key ideas, 85, 87, 88, 166
time frames, 45–46, 50–53; key ideas, 51, 53, 65
timing, of information, 137–138
Total Football, 150, 161
Tractor (Javier Zanetti), 99–102, 161
traffic. *See* roads and traffic
transfer fees, 92
transparency, 121, 135–139; definitions and key ideas, 136, 139, 146, 166
transportation planning and funding, 28–29, 93–94, 163
travel-related energy consumption, 113, 115–116, 117
trees, 119–120
Trinidad and Tobago, in 2006 World Cup, 77n27
triple bottom line accounting, 67n4
turbulent flow, 21–22n15
Tutu, Desmond, 1n1
Twitter, 14n30

uncertainty/unpredictability, 46, 61–65, 83, 107, 163; of climate-change impacts, 86n41, 120n2; key ideas, 62, 64, 65; soccer analogies, 61–62, 64, 75n25. *See also* adaptability; resilience
United Nations: definition of sustainability, 5, 66–67; Gender Inequality Index, 157n22; sustainable human population estimates, 103n26; World Commission on Environment and Development (Brundtland Commission), 66–67n3
university endowments, 137
University of California, 136–137
urban planning, 27–29, 49. *See also* transportation planning
Uruguay: the 1950 World Cup, 107–110, 111, 161
U.S.: female soccer fans in, 157n23; Major League Soccer and its innovations, 142–144, 145; U.S. World Cup teams, 12n25, 21n14, 45n1
U.S. Environmental Protection Agency, 71n13

Vargas, Getúlio, 108n32, 158–159
vegetarianism, 58n20
Vieira, Patrick, 7n15
Villas-Boas, André, 34n40
vision, 148, 149–153; definitions and key ideas, 151, 153, 164, 166
voting rights, for women, 157, 158, 160, 161

Wangchuck, Jigme Singye, 41–42
war, 67, 103, 141; climate change and, 103n25; Mozambique civil war, 16, 67, 90, 91, 148–149; soccer and, 4, 72n18
waste, human, managing, 46, 122
waste, reducing, 127. *See also* closing loops; inputs and outputs
water quality, 2, 23–24, 119
water-supply access and management, 17, 45–46; in Maputo, 45–46, 119, 120, 121, 122, 148
water use, for agriculture, 112n40
Weisman, Alan, 134n36
well-being. *See* happiness
West Ham United, 92, 135–136, 139, 153
what-if analysis. *See* counterfactual analysis
White Wilderness (film), 102
why questions. *See* five whys
Wilkins, Ray, 34n40
wind energy, 123–125

women's rights: in Brazil, 157, 158. *See also* gender equality
women's soccer, 156–157, 161, 163–164
World Cup soccer: African teams in, 47–48; Brazil's dominance, 111; vs. Champions League, 25–26; coach and manager behavior and influence, 30–32, 100–101; energy-efficient stadium designs, 112–113, 114; female players, 156, 157, 163; 1934, 72; 1950, 107–110, 111, 161; 1954, 59; 1958, 54–55, 73, 111; 1962, 54, 55, 72, 90; 1966, 7, 8, 46, 89–90, 91, 106n29; 1970 Soccer War, 4n8; 1974, 69–70, 72, 73, 150; 1978, 72; 1982, 4n12, 90n5; 1986, 100n20; 1990, 47–48, 145; 1994, 72n16, 73n22, 100n20; 1998, 7n15, 84, 90n5; number of tournament slots, 90; politics and, 69–70, 72; rules and rule changes, 144–145, 155; scoring averages, 155; suicide and, 107, 108; team participation in the early years, 110n37; 2006, 77n27, 83–85, 86–87, 97n16; 2007, 163; 2010, 20–22, 24, 83, 100–102, 135, 152–153, 157; 2014, 21–22n15, 112–113, 151n12; 2018 and 2022, 35n43, 90n5; U.S. teams, 12n25, 21n14, 45n1; war and, 72n18
World Player of the Year, 149n5, 152, 163

Zanetti, Javier, 99–102, 161
Ze (José Roberto Gomes Santana), 18–20
Zidane: A 21st Century Portrait (film), 84n38
Zidane, Zinedine, 84–85, 86–87, 98, 131, 132n30